Acoustics: A Unified Technical Glossary

From Wave Physics to Audio Engineering, Measurement, Modeling, and Perception

Edited by Jeffrey D. Iverson

BRIGHTFIELD
PRESS LLC

Table of Contents

Table of Contents.. 1
Introduction.. 2
Glossary Terms..3
Conclusion.. 196

and signal synchronization. They remain fundamental tools in audio engineering.

Acoustic Description — A structured explanation of a sound field, system, or environment using measurable acoustic parameters. It may include frequency response, reverberation time, directivity, and noise levels. Acoustic descriptions support design decisions and performance evaluations. They provide a common language for engineers, architects, and researchers.

Acoustic Energy — The mechanical energy carried by sound waves as they propagate through a medium. It depends on pressure variations and particle velocity. Acoustic energy can be concentrated, dispersed, absorbed, or reflected depending on the environment. Understanding energy flow is central to noise control and system design.

Acoustic Excitation — The process of driving a system with sound or vibration to study its response. Excitation may be impulsive, continuous, or swept across frequencies. It reveals resonances, damping characteristics, and structural behavior. Engineers use controlled excitation to characterize materials, rooms, and mechanical systems.

Acoustic Generator — A device that produces sound waves for testing, measurement, or signaling. Generators may use mechanical, electrical, or digital methods to create tones, noise, or complex signals. They support calibration, research, and system alignment. Precision generators are essential in laboratory acoustics.

Acoustic Holography — A technique that reconstructs a sound field in space using arrays of microphones and mathematical modeling. It allows visualization of sound sources, wave patterns, and radiation characteristics. Acoustic holography is used in noise diagnostics, product design, and research. It provides detailed spatial insight beyond traditional measurements.

Acoustic Impedance — A measure of the opposition a medium or system presents to sound propagation. It combines resistance, reactance, and mass effects into a single complex quantity. Impedance matching is crucial for efficient energy transfer between components. It underpins the behavior of resonators, transducers, and ducts.

Acoustic Impedance Matching Layer — A layer designed to improve energy transfer between two media with different acoustic impedances. It reduces reflections and increases transmission efficiency. Matching layers are common in ultrasonics, loudspeakers, and sensing devices. Their design depends on frequency, material properties, and application.

Acoustic Intensity Vector — A directional measure of acoustic energy flow, combining magnitude and direction. It is derived from sound pressure and particle velocity. Intensity vectors reveal how sound propagates through space and interact with boundaries. They are used in noise source identification and sound field analysis.

Acoustic Inversion — A computational technique that estimates sound sources or medium properties from measured acoustic data. It works backward from effects to causes, often using complex mathematical models. Acoustic inversion supports geophysics, medical imaging, and noise diagnostics. It is central to many modern acoustic reconstruction methods.

Acoustic Levitation — The suspension of small objects using standing sound waves that create stable pressure nodes. Levitation relies on precise control of acoustic fields and high-intensity sound. It enables contact-free manipulation in research and manufacturing. Applications include materials science, pharmaceuticals, and micro-assembly.

Acoustic Mach Meter — An instrument that measures the ratio of flow velocity to the speed of sound in a medium. It is used in aeroacoustics and high-speed flow studies. The Mach number influences shock formation, noise generation, and wave behavior. Accurate measurement is essential in aerodynamic and acoustic analysis.

Acoustic Mass — A property representing the inertial component of an acoustic system, analogous to inductance in electrical circuits. It describes how a volume of air resists acceleration when driven by sound pressure. Acoustic mass contributes to resonance behavior in ducts, ports, and cavities. It interacts with compliance and resistance to shape system response.

Acoustic Material — Any material specifically designed or selected for its sound-absorbing, reflecting, or diffusing properties. Acoustic materials shape the behavior of sound in rooms, enclosures, and devices. Their performance varies with frequency, thickness, and structure. They are essential tools in architectural and engineering acoustics.

Acoustic Metamaterial — An engineered material with structure-dependent acoustic properties not found in natural substances. Metamaterials can bend, block, or manipulate sound in unusual ways, including negative refraction. They enable advanced noise control, cloaking, and wave-shaping applications. Research continues to expand their capabilities.

Acoustic Modal Analysis — The study of how enclosed spaces or structures vibrate in discrete patterns called modes. Modal analysis identifies resonant frequencies, mode shapes, and damping characteristics. It is crucial in room acoustics, structural acoustics, and product design. Understanding modes helps control coloration, standing waves, and unwanted resonances.

Acoustic Power — The total sound energy emitted by a source per unit time. Unlike sound pressure, acoustic power is independent of distance and environment. It provides a standardized measure for comparing noise emissions. Power ratings are essential in product testing and regulatory compliance.

Acoustic Radiation Impedance — The effective impedance a vibrating surface experiences as it radiates sound into a medium. It includes both resistive and reactive components. Radiation impedance influences efficiency, bandwidth, and loading of transducers. It is fundamental in loudspeaker and ultrasonic design.

Acoustic Radiation Pressure — The steady force exerted by sound waves on a surface due to momentum transfer. It becomes significant at high intensities and in standing wave fields. Radiation pressure enables acoustic levitation and particle manipulation. It also affects transducer behavior and nonlinear acoustic phenomena.

Acoustic Reactance — The component of acoustic impedance that stores, rather than dissipates, energy. It arises from mass-like and spring-like behavior in acoustic systems. Reactance determines phase relationships between pressure and particle velocity. It plays a key role in resonance and system tuning.

Acoustic Reciprocity — A principle stating that the acoustic response between two points is the same regardless of which point acts as the source or receiver. Reciprocity simplifies measurement and modeling by allowing reversed configurations. It is widely used in transducer calibration and room acoustics. The principle holds under linear, time-invariant conditions.

Acoustic Refraction — The bending of sound waves as they pass through regions with different sound speeds, such as temperature gradients or layered media. Refraction alters propagation paths, focusing or dispersing sound. It affects outdoor sound propagation, underwater acoustics, and atmospheric studies. Understanding refraction improves prediction of long-range sound behavior.

Acoustic Resistance — The component of acoustic impedance that dissipates sound energy as heat. It represents true energy loss in a system. Acoustic resistance influences damping, absorption, and overall system efficiency. It works alongside mass and reactance to define acoustic behavior.

Acoustic Scattering — The redirection of sound waves when they encounter objects, surfaces, or irregularities in a medium. Scattering alters the distribution of acoustic energy and contributes to diffusion, coloration, and spatial complexity in a sound field. It is influenced by object size, shape, and material properties relative to wavelength. Engineers study scattering to improve room acoustics, noise control, and sonar performance.

Acoustic Stiffness — A measure of how resistant a system is to deformation when subjected to sound pressure. It represents the "spring-like" component of acoustic impedance. High stiffness raises resonant frequencies and reduces low-frequency response. Acoustic stiffness plays a key role in enclosure design, transducer behavior, and resonant cavity modeling.

Acoustic Streaming — A steady fluid flow generated by the absorption of high-intensity sound waves. This nonlinear effect occurs in gases and liquids and can move particles or influence heat transfer. Acoustic streaming is important in ultrasonics, microfluidics, and medical applications. It demonstrates how sound can exert mechanical influence beyond simple vibration.

Acoustic Tomography — A technique that reconstructs the properties of a medium by analyzing how sound waves travel through it. Variations in travel time, attenuation, or frequency content reveal temperature, density, or structural differences. Acoustic tomography is used in oceanography, atmospheric studies, and medical imaging. It provides a noninvasive way to "see" inside complex environments.

Acoustic Transfer Function — A mathematical description of how an acoustic system modifies sound from input to output. It captures frequency-dependent changes in amplitude and phase. Transfer functions are essential for system tuning, room correction, and loudspeaker design. They provide a foundation for predictive modeling and measurement analysis.

Acoustic Velocity — The speed at which sound waves propagate through a medium, determined by its elasticity and density. Acoustic velocity varies with temperature, pressure, and material composition. It influences refraction, resonance, and wave behavior in all acoustic systems. Accurate velocity values are essential for modeling and measurement.

Acoustic Vibration — The oscillatory motion of particles in a medium caused by sound waves. This motion transfers energy through air, liquids, or solids. Acoustic vibration underlies all sound propagation and interacts with structures, materials, and boundaries. Understanding vibration is fundamental to noise control and transducer design.

Acoustic Wave — A mechanical disturbance that propagates through a medium via pressure and particle motion. Acoustic waves can be longitudinal, transverse, or complex depending on the environment. Their behavior is shaped by frequency, wavelength, and boundary conditions. Acoustic waves form the basis of all sound phenomena.

Acoustics — The scientific study of sound, including its generation, transmission, measurement, and perception. Acoustics spans physics, engineering, architecture, biology, and psychology. It encompasses everything from wave behavior to hearing science and audio technology. The field provides the foundation for understanding and shaping sound in the world.

Active Circuitry — Electronic components that require power to operate and can amplify, filter, or modify audio signals. Active circuits use transistors, op-amps, or digital processors to achieve precise control. They enable advanced equalization, mixing, and signal routing. Active designs offer flexibility and performance beyond passive components.

Active Crossover — An electronic filter network that divides an audio signal into frequency bands before amplification. Each band is sent to a dedicated amplifier and loudspeaker driver. Active crossovers allow precise control of crossover points, slopes, and phase alignment. They improve system efficiency, clarity, and dynamic performance.

ADAT (Alesis Digital Audio Tape) — A multichannel digital recording format introduced by Alesis that allowed eight tracks of digital audio to be recorded onto S-VHS tape. ADAT revolutionized project-studio workflows by offering affordable, synchronized multitrack recording, with multiple machines linkable for up to 128 tracks. Its optical digital interface, still widely used today, became a standard for transferring eight channels of digital audio over a single fiber-optic cable. Although the tape-based system is now obsolete, the ADAT optical format remains a key connection method in modern digital audio equipment.

Adjoint Method — A computational technique used to efficiently calculate sensitivities in complex acoustic models. It allows engineers to determine how changes in parameters affect system behavior without rerunning full simulations. The adjoint method is widely used in optimization, inverse problems, and numerical acoustics. It enables high-precision modeling with reduced computational cost.

Aeroacoustics — The study of sound generated by airflow and fluid motion. It examines noise from turbulence, vortices, jets, and aerodynamic surfaces. Aeroacoustics is essential in aviation, automotive design, HVAC engineering, and environmental noise control. The field blends fluid dynamics with acoustic theory to understand and reduce flow-induced noise.

AES (Audio Engineering Society) — A professional organization dedicated to advancing audio science, engineering, and practice. The AES develops widely used technical standards, including digital-audio formats, connector specifications, and measurement protocols. It hosts international conventions, publishes peer-reviewed research, and supports collaboration across recording, acoustics, broadcast, and product design. AES standards and publications shape much of modern audio technology and ensure interoperability across equipment and workflows.

AF (Audio Frequency) — The range of frequencies that fall within the limits of human hearing, typically from 20 Hz to 20 kHz. AF signals represent the actual sound content carried through microphones, mixers, amplifiers, and loudspeakers. This range contrasts with RF (radio frequency), which is used for wireless transmission rather than audible sound. Understanding AF behavior is essential for system design, signal routing, and maintaining clarity across the audible spectrum.

AFL (After-Fader Listen) — A monitoring mode on a mixing console that lets the engineer listen to a channel *after* its fader and processing have been applied. AFL reflects the channel's contribution to the mix, including EQ, dynamics, and level settings. It is typically routed to the control-room or headphone bus without affecting the main output. AFL is essential for checking balance, troubleshooting mix issues, and verifying how a channel sits within the overall program.

A-Frequency Weighting — A standardized filter that adjusts sound measurements to approximate human hearing sensitivity at low levels. It emphasizes mid-frequencies and attenuates very low and very high frequencies. A-weighting is widely used in noise regulations and environmental assessments. It provides a consistent basis for comparing sound levels.

Airborne Sound — Sound that travels through the air rather than through solids or structures. Airborne sound includes speech, music, and environmental noise. Controlling airborne sound involves absorption, barriers, and insulation strategies. It contrasts with structure-borne sound, which travels through materials.

Airflow Resistance — A measure of how much a material resists the passage of air through it. It strongly influences a material's sound absorption characteristics, especially at mid- and high frequencies. Higher resistance generally increases absorption up to an optimal point. It is a key parameter in porous absorber design.

Airflow Resistivity — The airflow resistance of a material normalized by its thickness. Resistivity provides a more fundamental measure of acoustic behavior than resistance alone. It determines how effectively porous materials dissipate sound energy. Accurate resistivity values are essential for modeling absorbers and insulation.

Alcons — A metric expressing the percentage of speech articulation lost due to poor acoustics. Higher Alcons values indicate reduced intelligibility. It is influenced by reverberation, background noise, and system clarity. Alcons helps evaluate sound reinforcement and room performance.

Aliasing — A distortion that occurs when a signal is sampled at too low a rate, causing high-frequency components to appear as false low-frequency artifacts. Aliasing degrades audio quality and can introduce harsh or unnatural tones. Anti-aliasing filters and proper sampling rates prevent the effect. It is a fundamental concept in digital audio.

AM (Amplitude Modulation) — A method of encoding information onto a carrier wave by varying its amplitude in proportion to the input signal. In audio and radio systems, AM allows sound to be transmitted over long distances using radio-frequency carriers. Although more susceptible to noise and interference than FM, AM remains widely used in broadcasting and communication. Its simplicity makes it easy to generate, demodulate, and analyze in both analog and digital contexts.

Ambience — The natural background sound or reverberant character of a space. Ambience contributes to a sense of environment, depth, and realism in recordings. Engineers capture or simulate ambience to enhance spatial impression. It plays a major role in film, music, and immersive audio.

Ambient Coefficient — A parameter describing how environmental conditions influence sound propagation or material behavior. It may relate to temperature, humidity, or background noise effects. The term appears in modeling and environmental acoustics. Its meaning depends on the specific application.

Ambient Noise — The background sound present in an environment, excluding intentional sources. Ambient noise may include traffic, HVAC systems, wind, or distant activity. It sets the baseline for audibility and affects communication and comfort. Measuring ambient noise is essential in environmental and architectural acoustics.

Ambient Sound — The overall sonic atmosphere of a location, including both natural and human-made sources. Ambient sound shapes perception, mood, and spatial awareness. It is used creatively in sound design and evaluated technically in noise studies. Ambient sound differs from foreground or intentional audio.

Amp — A common abbreviation for "amplifier," referring to a device that increases the power of an audio signal. The term is widely used in music, recording, and sound reinforcement. Amps vary from small headphone units to large concert systems. Their design influences tone, clarity, and dynamic performance.

Ampere — The SI unit of electric current, representing the flow of one coulomb of charge per second. Amperes determine how much electrical power a device can draw or deliver. In audio systems, current capacity affects amplifier performance and loudspeaker control. Proper current handling ensures reliability and safety.

Amplifier — A device that increases the amplitude of an audio signal while preserving its waveform. Amplifiers power loudspeakers, drive headphones, and support signal routing Their performance is defined by gain, distortion, noise, and headroom. Amplifiers are central to all audio reproduction systems.

Amplifier Line — A signal path or connection dedicated to carrying amplified audio. It typically operates at higher voltage or power levels than line-level signals. Proper routing and shielding prevent interference and signal loss. Amplifier lines are integral to sound reinforcement and distributed audio systems.

Amplifier Output — The electrical signal delivered by an amplifier to a loudspeaker or downstream device. Output characteristics include voltage, current, impedance, and distortion. Matching amplifier output to speaker requirements ensures optimal performance. Output limits define system headroom and maximum loudness.

Amplitude — The magnitude of a sound wave's pressure variation or an electrical signal's voltage. Amplitude determines perceived loudness and influences dynamic expression. It is measured in decibels for sound and volts for electrical signals. Controlling amplitude is fundamental to recording, mixing, and system design.

Amplitude Distortion — A form of distortion where the output amplitude does not follow the input amplitude proportionally. It alters the waveform shape and introduces harmonic content. Excessive amplitude distortion reduces clarity and fidelity. High-quality systems minimize this effect through linear design.

Amplitude Shading — A technique used in loudspeaker arrays to adjust the amplitude of individual elements for improved coverage and directivity. Shading reduces lobing and smooths the overall radiation pattern. It is essential in line arrays and distributed systems. Proper shading enhances clarity and uniformity across an audience area.

Anacoustic Zone — A region in which reflections are so minimal that the sound field behaves as if it were free of boundaries. This zone is typically created in highly absorptive environments such as anechoic chambers. Within it, sound propagates without interference from surfaces, allowing precise measurements. It is essential for research, calibration, and transducer testing.

Analog — A continuous electrical or mechanical representation of a sound signal. Analog systems preserve subtle variations in amplitude and frequency, giving them a characteristic warmth and naturalness. They are susceptible to noise, distortion, and degradation over time. Despite digital dominance, analog technology remains valued in audio production and instrumentation.

Anechoic — Describing an environment or surface designed to eliminate sound reflections. Anechoic conditions simulate free-field propagation, allowing accurate measurement of sound sources. Such spaces rely on deep absorptive wedges and isolated construction. They are used in research, product testing, and psychoacoustic studies.

Anechoic Chamber — A specialized room engineered to absorb sound reflections across a wide frequency range. Its walls, ceiling, and floor are lined with wedge-shaped absorbers, and the structure is isolated from external noise. The chamber creates a near-perfect free-field environment. It is used for loudspeaker testing, microphone calibration, and acoustic research.

Anechoic Room — A room designed to minimize reflections but not necessarily to the extreme standards of a full anechoic chamber. It may provide free-field conditions above certain frequencies. Anechoic rooms support product testing, speech research, and controlled listening experiments. They balance practicality with acoustic precision.

ANSI (American National Standards Institute) — A nonprofit organization that oversees the development and coordination of voluntary technical standards in the United States. In acoustics and audio, ANSI approves measurement protocols, terminology standards, and performance specifications that ensure consistency across laboratories, manufacturers, and regulatory bodies. ANSI standards cover sound-level meters, noise measurement methods, hearing protection, and digital-audio formats. Its role is to unify practices across industries so that measurements and equipment remain reliable, comparable, and interoperable.

Antenna — A device that transmits or receives electromagnetic signals, including those used in wireless audio systems. Antennas convert electrical energy into radio waves and vice versa. Their design affects range, clarity, and interference resistance. They are essential in wireless microphones, communication systems, and RF-based monitoring.

Antenna Splitter — A device that distributes the signal from one antenna to multiple receivers. It allows several wireless channels to share a common antenna system without interference. Splitters reduce clutter and improve RF reliability in complex setups. They are widely used in broadcast, theater, and live sound.

Anti-Aliasing Filter — A low-pass filter applied before digital sampling to remove frequencies above half the sampling rate. Without it, high-frequency content would fold into the audible band as aliasing. The filter ensures accurate digital representation of analog signals. It is fundamental to all digital audio systems.

Anvil — One of the three small bones (ossicles) in the middle ear, formally called the incus. It transmits vibrations from the malleus to the stapes, helping amplify sound before it reaches the inner ear. The anvil plays a crucial role in hearing sensitivity. Its mechanical properties influence auditory perception.

Apparent Sound Reduction Index — A field-measured rating of how effectively a building element reduces airborne sound transmission. Unlike laboratory values, it accounts for real-world flanking paths and construction variability. It provides a more realistic assessment of acoustic performance. Designers use it to evaluate walls, floors, and partitions.

Arithmetic Mean Sound Pressure L — A statistical measure representing the simple average of sound pressure levels over time or across positions. It does not account for the logarithmic nature of decibels, so it is used sparingly in professional acoustics. The metric can provide rough comparisons in controlled contexts. More accurate measures typically use energy averaging.

Articulation — A measure of how clearly speech sounds are transmitted and understood in a given environment. It depends on reverberation, background noise, and system clarity. High articulation supports effective communication in classrooms, meeting rooms, and public spaces. It is a key metric in speech intelligibility studies.

Artificial Reverberation — Reverberation created electronically or digitally to simulate the acoustic character of a space. It enhances depth, realism, and spatial impression in recordings and sound reinforcement. Modern algorithms model early reflections, decay patterns, and diffusion. Artificial reverb is a core tool in audio production.

ASA (Acoustical Society of America) — A scientific organization dedicated to advancing the study and application of acoustics across research, engineering, medicine, architecture, and environmental science. The ASA publishes leading journals, develops technical standards, and hosts conferences that bring together experts from all acoustic disciplines. Its work supports progress in hearing science, noise control, underwater acoustics, architectural acoustics, and audio technology. ASA standards and research shape best practices throughout the global acoustics community.

ASCII (American Standard Code for Information Interchange) — A character-encoding system that represents letters, numbers, punctuation, and control commands using 7-bit binary values. ASCII provides a universal way for computers, digital audio devices, and communication systems to exchange text reliably. It underlies file formats, metadata structures, and communication protocols used throughout audio engineering. Despite newer Unicode standards, ASCII remains a foundational encoding scheme for digital systems.

ASHRAE (American Society of Heating, Refrigerating and Air-Conditioning Engineers) — A global professional organization that develops standards, research, and best practices for HVAC, building systems, indoor air quality, and energy efficiency. In acoustics, ASHRAE standards influence background-noise criteria, mechanical-system design, and airflow-related sound levels in occupied spaces. Its handbooks and guidelines help engineers balance thermal comfort, ventilation, and acoustic performance in buildings. ASHRAE's work shapes modern architectural design by integrating environmental control with occupant comfort and system reliability.

Assign — A term used in audio routing to indicate sending a signal to a specific bus, group, or output. Assigning allows flexible control over mixing, processing, and distribution. It is central to console operation and digital audio workflows. Clear assignment ensures organized and efficient signal flow.

Assistive Listening System — A technology that improves speech audibility for individuals with hearing difficulties. Systems may use FM, infrared, induction loops, or digital transmission. They deliver a clean, direct signal that bypasses room acoustics and background noise. Assistive listening is essential in public venues and accessibility compliance.

Attack — The initial portion of a sound or audio signal, describing how quickly it reaches full amplitude. Attack characteristics influence perceived sharpness, clarity, and impact. In audio processing, attack settings control how compressors and other dynamics tools respond. It is a key parameter in shaping sound.

Attenuate — To reduce the amplitude or intensity of a sound or signal. Attenuation may occur through distance, absorption, filtering, or electronic control. It helps manage levels, prevent overload, and shape tonal balance. Attenuation is fundamental to both acoustics and audio engineering.

Attenuation Coefficient — A measure of how quickly sound energy decreases as it travels through a medium. It accounts for absorption, scattering, and other loss mechanisms. Higher coefficients indicate stronger attenuation. The value is essential in modeling materials, ducts, and environmental propagation.

Attenuation Constant — A parameter describing the exponential decay of sound amplitude in a medium or waveguide. It is used in mathematical models of acoustic transmission. The constant reflects material properties, frequency, and boundary conditions. It helps predict signal loss over distance.

Attenuation Rate — The rate at which sound level decreases with distance or through a material. It may be expressed in decibels per meter or per octave band. Attenuation rate guides system design, noise control, and material selection. It varies with frequency and environmental conditions.

Attenuator — A device that reduces signal level without significantly altering its frequency content. Attenuators protect equipment, manage gain structure, and support calibration. They may be fixed, variable, or switchable. In acoustics, attenuators also refer to duct silencers that reduce airflow noise.

Audibility — The ability of a sound to be heard by a listener under specific conditions. Audibility depends on sound level, frequency, masking, and individual hearing sensitivity. It is central to noise assessment and communication design. Thresholds of audibility vary widely across listeners.

Audible Frequency Range — The range of frequencies that the average human ear can perceive, typically from 20 Hz to 20 kHz. Sensitivity varies across this range, with greatest acuity in the mid-frequencies. Age, exposure, and health influence individual limits. This range defines the foundation of audio system design.

Audio — Any sound or signal intended for recording, reproduction, communication, or analysis. Audio encompasses speech, music, effects, and environmental sound. It involves both acoustic and electronic processes. The field spans engineering, production, and scientific research.

Audio Chain — The sequence of devices and processes through which an audio signal passes. A typical chain may include microphones, processors, mixers, amplifiers, and loudspeakers. Each stage affects quality, noise, and dynamic range. A well-designed chain preserves clarity and fidelity.

Audio Frequency — Frequencies within the range of human hearing, generally 20 Hz to 20 kHz. Audio systems are designed to capture, process, and reproduce this range accurately. Different applications emphasize different portions of the spectrum. Audio frequency behavior shapes timbre, intelligibility, and spatial impression.

Audio Spectrum — The distribution of audio energy across frequencies. The spectrum reveals tonal balance, harmonic content, and noise characteristics. Engineers analyze the spectrum to diagnose issues and shape sound. It is fundamental to mixing, mastering, and acoustic measurement.

Audio Spectrum Analyzer — A tool that displays the amplitude of audio frequencies in real time. It helps identify peaks, resonances, and tonal imbalances. Spectrum analyzers are used in system tuning, troubleshooting, and production. They provide visual insight into complex audio signals.

Audiogram — A chart showing an individual's hearing thresholds across frequencies. Audiograms reveal hearing loss patterns and guide clinical assessment. They are essential in audiology, occupational health, and hearing-aid fitting. The chart provides a clear picture of auditory sensitivity.

Audiophile — A person dedicated to high-fidelity audio reproduction and critical listening. Audiophiles value precision, clarity, and naturalness in sound systems. Their interests span equipment design, room acoustics, and recording quality. The term reflects both technical knowledge and passion for sound.

Auditory Area — A region of the brain involved in processing sound information. It includes structures responsible for frequency analysis, localization, and pattern recognition. The auditory area integrates sensory input into meaningful perception. Its function underlies speech understanding and musical appreciation.

Auditory Cortex — The part of the cerebral cortex responsible for interpreting complex sound patterns. It processes pitch, timbre, spatial cues, and temporal structure. The auditory cortex plays a central role in speech, music, and environmental awareness. Its activity reflects both sensory input and learned experience.

Auditory Sensation Area — The region within the auditory system where sound is first perceived as a conscious sensation. It integrates basic acoustic features such as loudness, pitch, and timbre before higher-level processing occurs. This area bridges raw sensory input and meaningful interpretation. Its function underlies the earliest stages of hearing perception.

Auditory System — The collection of anatomical structures responsible for detecting, transmitting, and interpreting sound. It includes the outer, middle, and inner ear, along with neural pathways and brain regions. The system converts mechanical vibrations into electrical signals and ultimately into perception. Its performance shapes communication, spatial awareness, and auditory experience.

Aural — Relating to the ear or the sense of hearing. The term is used to describe listening experiences, perceptual qualities, and hearing-based evaluations. Aural assessments complement objective measurements in acoustics. They help capture how sound is actually perceived by listeners.

Automatic Mixer — A device or system that automatically adjusts audio levels based on signal activity. It reduces background noise and prevents overlapping speech by opening and closing channels as needed. Automatic mixers are widely used in conferencing, broadcast, and live sound. They improve clarity and reduce operator workload.

Aux- — A shorthand prefix for "auxiliary," commonly used in audio routing. It refers to secondary signal paths used for effects, monitoring, or external processing. Aux paths provide flexibility in mixing and system design. They are essential in both analog and digital workflows.

Auxiliary Send — A console or interface output that routes a portion of a channel's signal to an external destination. It may feed effects processors, monitor mixes, or recording devices. Sends can be pre- or post-fader depending on the application. They offer powerful control over signal distribution.

Average Room Absorption Coeffici — A shortened form of "average room absorption coefficient," representing the mean absorption of all surfaces in a room. It helps estimate reverberation time and overall acoustic behavior. The value is derived from material properties and surface areas. It is a key parameter in room acoustics calculations.

Average Sound Pressure Level — A time-averaged measure of sound pressure expressed in decibels. It smooths short-term fluctuations to represent typical exposure or environmental conditions. This metric is used in noise assessments, monitoring, and compliance testing. It provides a stable reference for comparing sound environments.

Average Sound Pressure Level In — A truncated form of a metric describing average sound pressure level in a specific context or location. It typically refers to measurements taken within a room, enclosure, or defined zone. The value helps assess acoustic performance or exposure. Full interpretation depends on the complete metric name.

A-Weighted Equivalent Continuous Sound Level — A long-term average sound level adjusted using the A-weighting curve to reflect human hearing sensitivity. It represents cumulative noise exposure over a defined period. This metric is widely used in environmental noise regulation and occupational safety. It provides a realistic measure of perceived loudness.

A-Weighted Sound Pressure Level — A sound pressure level measurement filtered through the A-weighting curve. It emphasizes mid-frequencies where human hearing is most sensitive. The metric is standard in noise assessments and compliance testing. It offers a consistent basis for comparing sound levels.

A-Weighting — A frequency weighting curve that approximates human hearing sensitivity at low sound levels. It reduces low-frequency and high-frequency content to match perceptual response. A-weighting is used in most environmental and occupational noise measurements. It provides a standardized way to quantify perceived loudness.

Axial Mode — A room resonance that occurs between two parallel surfaces, such as opposite walls. Axial modes strongly influence low-frequency behavior and can cause peaks or nulls in response. They are the most dominant room modes due to their simple geometry. Managing axial modes is essential for accurate listening environments.

Axis — A reference line used to describe directionality or orientation in acoustics. In loudspeakers, the on-axis direction is directly in front of the driver and typically yields the flattest response. Off-axis behavior reveals dispersion and tonal changes. Axis definitions support measurement and system alignment.

Backcountry — A term referring to remote outdoor environments, sometimes used in environmental acoustics. Backcountry soundscapes are dominated by natural sources and low ambient noise. These areas are studied for conservation, recreation planning, and noise impact assessment. They provide benchmarks for natural acoustic conditions.

Background Noise — The ambient sound present in an environment, excluding intentional sources. It may include HVAC systems, traffic, wind, or distant activity. Background noise affects communication, comfort, and measurement accuracy. Controlling it is essential in critical listening and speech environments.

Background Noise Level — A quantified measure of the background noise present in a space. It is typically expressed in decibels and may use frequency weighting. This level sets the baseline for audibility and speech intelligibility. Designers use it to evaluate room performance and noise control needs.

Baffle — A surface or structure used to control the radiation or propagation of sound. In loudspeakers, baffles support drivers and influence dispersion. In acoustics, baffles can block, redirect, or absorb sound. Their design affects clarity, coverage, and tonal balance.

Baffle Step — A phenomenon where a loudspeaker's output transitions from half-space to full-space radiation. This causes a drop in low-frequency response as wavelengths exceed baffle dimensions. Designers compensate for baffle step through crossover design or equalization. It is a key factor in loudspeaker voicing.

Balance — The relative level relationship between two audio channels or frequency regions. Proper balance ensures natural imaging and tonal neutrality. It is adjusted during mixing, playback, and system tuning. Good balance enhances clarity and listener comfort.

Balance Control — A user interface element that adjusts the left-right level distribution in a stereo system. It compensates for room asymmetry, listener position, or recording imbalance. The control helps maintain proper stereo imaging. It is common in consumer and professional audio equipment.

Balanced — A method of audio signal transmission that uses two opposite-polarity conductors to reduce noise and interference. Balanced lines offer superior rejection of hum and RF contamination. They are standard in professional audio systems. Balanced connections improve signal integrity over long cable runs.

Band — A defined range of frequencies within the audio spectrum. Bands are used in filtering, equalization, and measurement. They help isolate specific tonal regions or noise components. Band definitions support precise control and analysis.

Band Pressure Level — The sound pressure level measured within a specific frequency band. It provides insight into spectral distribution and tonal balance. Engineers use band levels to diagnose issues and evaluate system performance. It is fundamental in noise analysis and room acoustics.

Bandpass Filter — A filter that allows a specific range of frequencies to pass while attenuating frequencies above and below it. Bandpass filters shape tonal character, isolate signals, and support measurement. They are used in audio processing, communication, and instrumentation. Their design determines bandwidth and center frequency.

Bandwidth — The width of a frequency range that a system, filter, or device can effectively transmit. Wider bandwidth supports greater detail and clarity, while narrow bandwidth isolates specific components. Bandwidth is central to audio quality, communication, and measurement. It influences system performance across many applications.

Bark Scale — A psychoacoustic frequency scale based on critical bands of human hearing. It reflects how the ear groups frequencies into perceptual units. The Bark scale supports loudness modeling, speech analysis, and audio coding. It provides a perceptually meaningful alternative to linear frequency scales.

Barriers — Structures designed to block or reduce sound transmission between spaces. Barriers rely on mass, stiffness, and airtight construction to impede sound. They are used in noise control, building design, and environmental acoustics. Effective barriers reduce unwanted intrusion and improve acoustic comfort.

Basilar Membrane — A key structure in the cochlea that converts mechanical vibrations into neural signals. It is tonotopically organized, with different regions responding to different frequencies. Movement of the basilar membrane stimulates hair cells, initiating auditory perception. Its function underlies pitch discrimination and hearing sensitivity.

Bass — The low-frequency region of the audio spectrum, typically below 250 Hz. Bass contributes weight, warmth, and power to sound. Its behavior is strongly influenced by room acoustics and loudspeaker design. Proper bass control enhances clarity and musical impact.

Bass Boost — An increase in low-frequency amplitude applied through equalization or circuitry. Bass boost enhances fullness and impact in playback systems. It compensates for small speakers or low-level listening. Excessive boost can cause distortion or masking.

Beacon — A signaling device that emits sound for navigation, warning, or identification. Acoustic beacons are used in aviation, marine environments, and emergency systems. Their design emphasizes reliability and audibility. They play a critical role in safety and communication.

Beam Pattern — The directional distribution of sound radiated by a loudspeaker or microphone. Beam patterns reveal how energy spreads across angles and frequencies. They influence coverage, clarity, and feedback control. Understanding beam patterns is essential in system design.

Beam Steering — The electronic or mechanical control of a loudspeaker array's radiation direction. Steering allows sound to be aimed without physically moving the system. It improves coverage, reduces reflections, and enhances intelligibility. Modern systems use digital signal processing for precise control.

Beam Width — The angular width of a loudspeaker's or microphone's main radiation lobe. It defines how broadly or narrowly sound is projected or captured. Beam width varies with frequency and device design. Proper beam width selection ensures consistent coverage and tonal balance.

Beamforming — A signal-processing technique that shapes and directs the output of a microphone or loudspeaker array. By adjusting timing and amplitude across elements, beamforming can focus sound in specific directions or suppress unwanted noise. It improves clarity, coverage, and intelligibility in complex environments. Modern systems use digital algorithms for precise, adaptive control.

Beaming — A phenomenon where a loudspeaker becomes increasingly directional at high frequencies. As wavelengths shorten relative to driver size, sound radiates in a narrow beam rather than spreading evenly. Beaming can cause uneven tonal balance across a listening area. Designers manage it through driver selection, waveguides, and crossover strategies.

Beats — A periodic fluctuation in amplitude created when two tones of slightly different frequencies interfere. The resulting "beat frequency" equals the difference between the two tones. Beats are used in tuning, psychoacoustic research, and modulation effects. They illustrate fundamental principles of wave interaction.

Behavioral Response and Habituation — A concept describing how organisms react to repeated sound exposure and gradually reduce their response over time. Habituation helps distinguish meaningful signals from background noise. In acoustics, it informs studies of wildlife, human perception, and noise tolerance. Understanding habituation guides environmental and psychological sound research.

Bel — A logarithmic unit used to express ratios of power or intensity, equal to ten decibels. The bel provides a convenient way to describe large dynamic ranges. Although rarely used directly, it forms the basis of the decibel scale. Its name honors Alexander Graham Bell.

Best Frequency — The frequency at which a particular auditory neuron or hair cell responds most strongly. Best frequency reflects the tonotopic organization of the auditory system. It plays a key role in pitch perception and frequency discrimination. Researchers use it to map auditory pathways.

Bi-Amped — A loudspeaker configuration in which separate amplifiers power different frequency bands, typically lows and highs. Bi-amping improves control, reduces distortion, and enhances dynamic performance. It allows precise crossover implementation and driver matching. Professional systems often use bi-amping for superior clarity.

Bias Voltage — A steady electrical voltage applied to a microphone or electronic component to optimize its operating point. In condenser microphones, bias voltage charges the diaphragm and backplate. Proper biasing ensures sensitivity, stability, and low distortion. It is essential for accurate signal capture.

Bidirectional Microphone — A microphone with a figure-eight pickup pattern, capturing sound equally from the front and rear while rejecting the sides. This pattern is useful for interviews, mid-side recording, and certain stereo techniques. Bidirectional mics offer strong side rejection and natural ambience capture. Their behavior depends on capsule design and placement.

Binaural — Relating to the use of two ears or two channels to create spatial hearing. Binaural techniques capture or simulate the cues the brain uses for localization. They enable immersive headphone experiences and realistic spatial audio. Binaural processing is central to psychoacoustics and 3D sound.

Binaural Room Impulse Response — A room impulse response recorded using a binaural microphone or dummy head. It captures spatial cues such as direction, reflections, and reverberation. BRIRs are used in virtual acoustics, auralization, and immersive audio design. They provide highly realistic spatial reproduction.

Bioacoustics — The study of sound production, perception, and communication in animals. It explores how species use sound for navigation, mating, territory, and survival. Bioacoustics supports conservation, behavioral research, and ecological monitoring. It bridges biology, acoustics, and environmental science.

Biot Theory — A theoretical framework describing sound propagation in porous materials. It accounts for interactions between the solid frame and the fluid within the pores. Biot theory underpins modern modeling of acoustic insulation, foams, and fibrous absorbers. It is essential for advanced material design.

Bit Depth — The number of bits used to represent each audio sample in digital systems. Higher bit depth increases dynamic range and reduces quantization noise. Common depths include 16-bit, 24-bit, and 32-bit float. Bit depth strongly influences perceived clarity and headroom.

Blanking Distance — The minimum distance required for a directional microphone or sensor to detect a target without interference from its own near-field effects. It ensures accurate measurement by avoiding self-generated noise or reflections. Blanking distance is important in sonar, radar, and specialized acoustic sensing. Its value depends on frequency and device design.

Block — A general term for a section of an audio system, signal chain, or processing module. Blocks represent functional units such as filters, mixers, or amplifiers. Block diagrams help visualize system architecture and signal flow. They support troubleshooting and design clarity.

Body-Pack — A compact wireless transmitter worn on the body, typically used with lavalier or headset microphones. Body-packs provide mobility and convenience in theater, broadcast, and live sound. They include controls for gain, frequency, and power. Proper placement ensures reliable performance and minimal noise.

Boomy — A descriptive term for excessive or uncontrolled low-frequency resonance. Boomy sound often results from room modes, poor placement, or inadequate damping. It reduces clarity and masks mid-frequency detail. Acoustic treatment and careful tuning help eliminate boominess.

Boost — An increase in amplitude applied to a specific frequency range or signal. Boosting can enhance presence, warmth, or clarity when used judiciously. Excessive boost may introduce distortion or imbalance. It is a fundamental tool in equalization and mixing.

Boundary Element Method — A numerical technique used to model acoustic fields by solving equations on the boundaries of a region rather than throughout its volume. BEM is efficient for problems involving complex surfaces and open spaces. It supports loudspeaker design, noise prediction, and scattering analysis. The method is widely used in computational acoustics.

Boundary Layer Noise — Noise generated by turbulent airflow near a surface, such as an aircraft fuselage or vehicle body. It contributes to aerodynamic noise and affects passenger comfort. Boundary layer noise increases with speed and surface roughness. Engineers study it to improve aerodynamic and acoustic performance.

Boundary Microphone — A microphone designed to operate on or near a reflective surface, using the boundary effect to improve sensitivity and reduce phase interference. Boundary mics provide clear, natural pickup in conference rooms, stages, and recording environments. They minimize comb filtering and capture wide coverage. Their low-profile design enhances placement flexibility.

Braided Shield — A woven metallic layer surrounding an audio cable to protect against electromagnetic interference. Braided shields offer strong coverage and durability. They reduce hum, RF noise, and signal contamination. High-quality cables rely on effective shielding for clean audio transmission.

Break — A term used to describe a sudden change in frequency response, system behavior, or mechanical operation. In loudspeakers, break-up modes occur when the diaphragm no longer moves uniformly. Breaks can introduce distortion or coloration. Designers work to minimize or control these effects.

Bridged Mono — A configuration where two amplifier channels are combined to drive a single load with increased power. Bridging doubles voltage swing and significantly boosts output. It is useful for subwoofers and high-demand applications. Proper impedance matching is essential to avoid overload.

Bright — A tonal descriptor indicating emphasis in the high-frequency range. Bright sound can enhance clarity and detail but may become harsh if excessive. It is influenced by equalization, transducer design, and room acoustics. Engineers adjust brightness to achieve balanced reproduction.

Broad-Band Noise — Noise that contains energy across a wide range of frequencies. Examples include white noise, pink noise, and environmental noise. Broad-band noise is used in testing, masking, and acoustic analysis. It provides a uniform stimulus for evaluating system response.

Broadcast — The transmission of audio or audiovisual content to a wide audience via radio, television, or digital platforms. Broadcast audio requires consistent quality, intelligibility, and reliability. It involves specialized equipment, standards, and workflows. Acoustics plays a major role in studio design and signal processing.

Bulk Modulus — A measure of a material's resistance to compression. In acoustics, it helps determine sound speed and wave behavior in fluids and solids. Higher bulk modulus indicates greater stiffness and faster sound propagation. It is fundamental to physical acoustics and material modeling.

Burst — A short, intense sound event characterized by rapid onset and decay. Bursts are used in testing to evaluate transient response and system dynamics. They also occur naturally in impacts, mechanical events, and certain speech sounds. Burst analysis reveals nonlinearities and time-domain behavior.

Buzz — An unwanted sound characterized by low-frequency or harmonic noise, often caused by electrical interference or mechanical vibration. Buzz can originate from grounding issues, loose components, or faulty equipment. It degrades audio quality and distracts listeners. Proper troubleshooting eliminates the source.

Byte — A digital information unit consisting of eight bits. In audio, bytes store sample values, metadata, and file structure. Byte depth influences resolution and dynamic range. Understanding bytes is essential for digital audio formats and processing.

Cable — A physical medium used to transmit audio, power, or data signals. Cable design affects noise rejection, impedance, durability, and signal integrity. Types include balanced, unbalanced, digital, and speaker cables. Proper cable selection ensures reliable system performance.

Cannon — A colloquial term for the XLR connector, originally manufactured by Cannon Electric. XLR connectors provide secure, balanced audio transmission. They are standard in professional audio for microphones, mixers, and interfaces. Their durability and locking mechanism ensure dependable connections.

Cans — A slang term for headphones, commonly used in studio and live-sound environments. Cans provide isolated monitoring for performers and engineers. Their design influences comfort, accuracy, and noise isolation. High-quality cans are essential for critical listening and production work.

Capacitor — An electronic component that stores electrical energy in an electric field. In audio circuits, capacitors are used for filtering, coupling, and stabilizing signals. Their value and type influence frequency response and noise performance. High-quality capacitors help maintain clarity and reduce distortion in sensitive audio paths.

Capstan — A rotating spindle used in tape machines to pull magnetic tape at a constant speed. The capstan works with a pinch roller to ensure stable transport and minimize wow and flutter. Its precision directly affects pitch accuracy and recording quality. Proper maintenance is essential for reliable tape performance.

Capsule — The core element of a microphone that converts sound waves into electrical signals. Capsules vary in design, including dynamic, condenser, and ribbon types, each with distinct characteristics. Their construction determines sensitivity, frequency response, and polar pattern. The capsule is the heart of microphone performance.

Cardioid — A microphone polar pattern shaped like a heart, capturing sound primarily from the front while rejecting the rear. Cardioid mics reduce feedback and background noise, making them ideal for live sound and studio use. Their directional behavior varies with frequency. Proper placement maximizes clarity and isolation.

Carrier — A high-frequency signal used to transmit audio or data in wireless and modulation systems. The audio signal modifies the carrier through amplitude, frequency, or phase changes. Carriers enable long-distance and interference-resistant communication. They are fundamental to wireless microphones and broadcast systems.

Cauchy Number — A dimensionless parameter comparing inertial forces to elastic forces in a medium. In acoustics, it helps characterize wave propagation in materials under stress. High Cauchy numbers indicate dominance of inertial effects. It is used in advanced modeling of structural and fluid acoustics.

Cavitation Threshold — The minimum sound pressure level required to create vapor bubbles in a liquid. Cavitation occurs under intense ultrasonic fields and can cause erosion or chemical effects. The threshold depends on frequency, temperature, and liquid properties. It is important in medical ultrasonics and industrial processing.

Cavity — An enclosed space that can resonate when excited by sound. Cavities amplify or attenuate specific frequencies depending on their dimensions. They appear in loudspeaker enclosures, architectural spaces, and mechanical systems. Understanding cavity behavior helps control resonance and coloration.

Ceiling Attenuation Class — A rating that measures how effectively a ceiling system reduces sound transmission between adjacent spaces. CAC values help evaluate acoustic privacy in offices and commercial buildings. Higher ratings indicate better sound isolation. Designers use CAC to select ceiling tiles and construction methods.

Cent — A unit used to measure small musical pitch intervals, with 100 cents equaling one semitone. Cents allow precise comparison of tuning differences. They are widely used in psychoacoustics, instrument design, and digital audio. The scale reflects human sensitivity to pitch variation.

Center Cluster — A loudspeaker arrangement placed at the center of a venue to provide even coverage. Center clusters reduce stereo imbalance and improve speech intelligibility. They are common in auditoriums, houses of worship, and multipurpose spaces. Proper design ensures consistent sound across the audience.

Channel — A discrete audio signal path used for recording, mixing, or playback. Channels may represent microphones, instruments, or speaker feeds. Each channel includes controls for level, processing, and routing. Multi-channel systems support complex mixes and immersive formats.

Channel Balance — The relative level relationship between left and right channels in a stereo system. Proper balance ensures accurate imaging and natural soundstage. Imbalance can result from equipment issues, room acoustics, or recording errors. Adjustments restore clarity and spatial accuracy.

Channel Strip — A vertical section of a mixing console or digital interface containing controls for a single audio channel. It typically includes gain, EQ, dynamics, routing, and fader controls. Channel strips streamline workflow and provide consistent access to essential functions. They are central to mixing and signal management.

Characteristic Impedance — The inherent impedance of a transmission line or medium that determines how signals propagate through it. Matching characteristic impedance minimizes reflections and signal loss. In acoustics, it applies to ducts, tubes, and waveguides. Proper impedance matching ensures efficient energy transfer.

Characteristic Impedance of the — A truncated form of a term referring to the characteristic impedance of a specific system or medium. Full interpretation requires the complete phrase. It typically relates to waveguides, cables, or acoustic ducts. The concept is essential for understanding transmission behavior.

Chorusing — An audio effect created by mixing a signal with delayed, pitch-modulated copies of itself. Chorusing adds richness, width, and movement to sound. It is widely used in music production for vocals, guitars, and synthesizers. The effect simulates multiple similar sources playing together.

Circuit — A closed path through which electrical current flows. Audio circuits process, amplify, filter, and route signals. Their design influences noise, distortion, and overall fidelity. High-quality circuits are essential for accurate audio reproduction.

Clip — A condition where an audio signal exceeds the maximum level a device can handle, causing waveform distortion. Clipping introduces harsh, unwanted harmonics and reduces clarity. It can damage equipment or speakers if severe. Proper gain staging prevents clipping.

Clip LED — An indicator light that signals when a device is approaching or reaching clipping. It helps engineers maintain proper levels and avoid distortion. Clip LEDs are found on mixers, interfaces, and amplifiers. Monitoring them ensures clean, controlled audio.

Clipping — The distortion that occurs when an amplifier or processor cannot reproduce peaks beyond its voltage or current limits. The waveform becomes flattened, introducing strong harmonics. Clipping reduces fidelity and can harm loudspeakers. Avoiding clipping is a core principle of gain management.

Clipping Level — The signal level at which a device begins to clip. It defines the usable dynamic range and headroom. Knowing the clipping level helps engineers set safe operating levels. Devices with higher clipping thresholds offer greater performance flexibility.

Close Pickup — A microphone technique where the mic is placed very near the sound source. Close pickup increases isolation, reduces room influence, and enhances presence. It is common in studio recording and live sound reinforcement. Proximity effects must be managed for natural results.

Cluster — A group of loudspeakers arranged to provide controlled coverage over a target area. Clusters improve uniformity and power handling in large venues. Their configuration affects directivity, phase interaction, and tonal balance. Proper design ensures consistent audience experience.

Coax Cable — A type of cable with a central conductor surrounded by insulation, shielding, and an outer jacket. Coaxial construction provides excellent RF shielding and consistent impedance. It is used for digital audio, antennas, and video transmission. Coax cables ensure reliable, low-noise signal transport.

Coaxial — A design in which two conductors share a common axis, typically referring to coaxial cables or coaxial loudspeakers. Coaxial speakers place drivers on the same axis for improved phase coherence and imaging. Coaxial cables provide stable impedance and strong shielding. The design enhances performance in compact systems.

Cochlea — The spiral-shaped organ in the inner ear responsible for converting sound vibrations into neural signals. It contains hair cells arranged tonotopically to detect different frequencies. The cochlea enables pitch perception, loudness discrimination, and auditory detail. Its function is central to human hearing.

Coherence — A measure of the correlation between two signals across frequency. High coherence indicates consistent phase and amplitude relationships. It is used in measurement, noise analysis, and system tuning. Low coherence suggests interference, noise, or system mismatch.

Coincidence Effect Critical Frequency — The frequency at which bending waves in a panel match the speed of airborne sound, causing increased sound transmission. At this critical frequency, sound insulation performance drops sharply. Designers mitigate the effect through material selection and damping. It is a key factor in building acoustics.

Coloration — A change in the natural timbre of sound caused by resonances, reflections, or frequency response irregularities. Coloration can make audio sound boxy, nasal, or harsh. It arises from room acoustics, equipment design, or microphone placement. Reducing coloration improves clarity and accuracy.

Comb Filter — A frequency response pattern created when a signal combines with a delayed version of itself. The result is a series of peaks and nulls resembling a comb. Comb filtering occurs in reflections, multi-mic setups, and digital processing. It can degrade clarity and tonal balance if uncontrolled.

Community Noise Equivalent Level — A long-term average noise metric used to evaluate environmental noise impact. It incorporates penalties for evening and nighttime noise to reflect human sensitivity. The metric supports urban planning, regulation, and environmental assessment. It provides a comprehensive view of community noise exposure.

Companding — A process that compresses a signal's dynamic range before transmission and expands it afterward. Companding reduces noise and improves signal-to-noise ratio in communication systems. It is used in wireless microphones, telephony, and analog recording. The technique preserves audio quality under limited bandwidth.

Compandor — A device that performs companding, combining compression and expansion in a single system. It improves noise performance in analog and wireless audio. Proper calibration ensures transparent operation. Compandors remain important in legacy and specialized communication systems.

Complex Impedance — A representation of impedance that includes both resistive and reactive components. It describes how systems store and dissipate energy across frequencies. Complex impedance is fundamental in modeling transducers, ducts, and acoustic materials. It guides design and matching for optimal performance.

Complex Tone — A sound composed of multiple frequency components, typically harmonics of a fundamental frequency. Complex tones form the basis of musical timbre and speech. Their structure influences perception of pitch, richness, and character. Understanding complex tones is central to psychoacoustics and audio synthesis.

Compression — A reduction in the dynamic range of an audio signal, making quiet sounds louder and loud sounds quieter. Compression helps control peaks, improve clarity, and maintain consistent levels. It is widely used in recording, broadcasting, and live sound. Proper settings preserve natural dynamics while preventing distortion.

Compression Driver — A type of high-frequency loudspeaker that uses a diaphragm and phase plug to project sound through a horn. Compression drivers achieve high efficiency and controlled directivity. They are essential in sound reinforcement systems where long-throw performance is required. Their design influences clarity, power handling, and tonal character.

Compressional Wave — A type of longitudinal wave in which particles in a medium oscillate parallel to the direction of propagation. Sound in air and liquids travels primarily as compressional waves. These waves alternate between compression and rarefaction. Their behavior defines fundamental acoustic properties such as speed and wavelength.

Compressor — An audio processor that automatically reduces signal peaks to control dynamic range. Compressors help maintain consistent levels, prevent overload, and shape tonal character. Parameters such as threshold, ratio, attack, and release determine behavior. They are indispensable tools in mixing and mastering.

Concave — Describing a surface that curves inward like the inside of a bowl. Concave surfaces can focus sound waves creating hotspots or reflections. They are used intentionally in some acoustic designs but can cause problems if uncontrolled. Proper treatment prevents unwanted focusing effects.

Condensor — A legacy or alternate spelling of "condenser," referring to a capacitor or condenser microphone. In audio, the term most often appears in older documentation. Modern usage favors "condenser." The meaning depends on context.

Condensor Microphone — A microphone that uses a charged diaphragm and backplate to convert sound into electrical signals. Condenser mics offer high sensitivity, wide frequency response, and detailed sound. They require external power such as phantom power or a battery. They are widely used in studios, broadcast, and critical recording.

Conductor — A material that allows electrical current to flow easily. In audio, conductors form the core of cables, connectors, and electronic circuits. Their resistance, purity, and construction influence signal integrity. High-quality conductors help reduce noise and maintain fidelity.

Cone — The diaphragm of a loudspeaker, typically shaped like a cone to move air efficiently. Cone materials affect stiffness, damping, and frequency response. Larger cones reproduce lower frequencies, while smaller ones handle mids and highs. Cone behavior strongly influences loudspeaker performance.

Connector — A device that joins audio cables or components, enabling signal transfer. Common types include XLR, TRS, RCA, and Speakon. Connectors affect reliability, noise rejection, and compatibility. Proper connectors ensure clean, stable audio connections.

Constant Voltage — A method of distributing audio signals over long distances using standardized voltage levels, often 70V or 100V systems. Constant-voltage systems simplify wiring and allow multiple speakers on a single line. They are common in commercial and public-address installations. Transformers match speaker loads to maintain efficiency.

Continuous Spectrum — A spectrum containing energy at all frequencies within a range, without discrete peaks. Examples include white noise and certain natural sounds. Continuous spectra are useful for testing and analysis. They reveal system behavior across the entire frequency band.

Convex — Describing a surface that curves outward like the exterior of a sphere. Convex surfaces scatter sound, promoting diffusion and reducing focused reflections. They are used in acoustic treatment to improve spatial uniformity. Proper placement enhances room clarity and naturalness.

Correlogram — A visual representation of the correlation between a signal and delayed versions of itself. Correlograms reveal periodicity, pitch cues, and temporal structure. They are used in psychoacoustics, speech analysis, and signal processing. The technique helps model how the auditory system detects patterns.

Crest Factor — The ratio of a signal's peak amplitude to its RMS level. High crest factors indicate dynamic, transient-rich signals, while low crest factors reflect compressed or steady signals. Crest factor influences headroom, distortion risk, and amplifier requirements. It is a key metric in system design.

Crimp — A mechanical method of attaching connectors to cables by compressing metal components around conductors. Proper crimping ensures secure electrical contact and strain relief. Poor crimps cause noise, dropouts, or failure. Professional tools and technique are essential for reliability.

Critical Band — A frequency region within which the ear integrates sound energy and experiences masking effects. Critical bands form the basis of many psychoacoustic models. They influence loudness perception, pitch discrimination, and audio coding. The concept helps explain how humans process complex sounds.

Critical Distance — The distance from a sound source at which direct sound and reverberant sound are equal in level. Beyond this point, clarity decreases and intelligibility suffers. Critical distance depends on room acoustics and source directivity. It guides microphone placement and sound system design.

Critical Frequency — The frequency at which a panel or structure becomes highly efficient at transmitting sound due to coincidence effects. At this point, airborne sound couples strongly with bending waves in the material. Sound insulation performance drops significantly. Designers mitigate this through damping and material selection.

Critical Ratio — The difference in decibels between a tone and the masking noise required for the tone to be just audible. It reflects the ear's ability to detect signals in noise. Critical ratio varies with frequency and listening conditions. It is important in communication, audio coding, and psychoacoustics.

Crossover — A filter network that divides an audio signal into frequency bands for separate loudspeaker drivers. Crossovers ensure each driver handles only the frequencies it is designed for. They may be passive, active, or digital. Proper crossover design improves clarity, efficiency, and tonal balance.

Crossover Frequency — The frequency at which a crossover divides the audio spectrum between drivers. It determines how lows, mids, and highs are distributed. Smooth transitions at the crossover frequency prevent phase issues and tonal gaps. Designers choose crossover points based on driver capabilities.

Crosstalk — Unwanted leakage of one audio signal into another channel. Crosstalk reduces stereo separation and can cause confusion or noise. It arises from poor shielding, circuit layout, or cable proximity. High-quality equipment minimizes crosstalk for clean, isolated signals.

Cue — A signal or instruction used to trigger an action in audio production or performance. Cues may indicate when to start playback, adjust levels, or execute effects. They help coordinate complex workflows in theater, broadcast, and live sound. Clear cues ensure smooth operation.

Current — The flow of electric charge through a conductor, measured in amperes. In audio systems, current determines power delivery to loudspeakers and affects amplifier performance. Insufficient current can cause distortion or failure. Proper current handling ensures reliable operation.

Cut — A reduction in amplitude applied to a specific frequency range or signal. Cutting helps remove unwanted resonances, noise, or harshness. It is a fundamental equalization technique. Effective cuts improve clarity and tonal balance.

Cutoff Frequency — The frequency at which a filter begins to significantly attenuate a signal. It defines the boundary between passband and stopband. Cutoff frequency shapes tonal character and system response. Precise control is essential in crossover and EQ design.

C-Weighted Peak Sound Level — A measurement of peak sound pressure using the C-weighting curve, which emphasizes low-frequency content. It is used in assessing impulsive or high-energy noise. C-weighted peak levels help evaluate hearing risk and regulatory compliance. The metric captures extreme sound events more accurately than A-weighting.

Cycles Per Second — An older term for frequency, now expressed in hertz (Hz). It describes how many times a waveform repeats each second. The term appears in legacy documentation and early audio literature. Modern usage favors the standardized SI unit.

Damp — To reduce the amplitude of vibrations or sound through absorption or resistance. Damping helps control resonances and prevent excessive ringing. It is applied in materials, enclosures, and mechanical systems. Effective damping improves clarity and stability.

Damping — The process of dissipating vibrational energy in a system. Damping reduces resonance peaks and shortens decay times. It is essential in loudspeaker design, room acoustics, and structural engineering. Proper damping enhances accuracy and reduces coloration.

DAT (Digital Audio Tape) — A digital recording format introduced by Sony that stores high-quality audio on small magnetic cassettes. DAT supports 16-bit recording at sample rates such as 44.1 kHz and 48 kHz, offering near-CD-quality fidelity in a compact, portable medium. It became a standard in professional studios for mixdowns, backups, and field recording thanks to its low noise, stable timecode, and reliable digital transfers. Although now largely obsolete, DAT played a major role in the transition from analog to digital production workflows.

Day-Night Average Sound Level — A 24-hour noise metric that applies penalties to evening and nighttime noise to reflect increased sensitivity. It is used in environmental noise regulation and urban planning. The metric provides a realistic assessment of community noise exposure. It guides policy and mitigation strategies.

dB (Decibel) — A logarithmic unit used to express ratios of power, pressure, or intensity in audio and acoustics. Because human hearing responds logarithmically, the dB scale provides a practical way to describe very large or very small changes in level. A decibel value always refers to a ratio, and its meaning depends on the reference used, such as dB SPL for sound pressure level or dBu and dBV for electrical signal levels. The dB scale simplifies comparisons, highlights perceptually meaningful differences, and is fundamental to measurement, system design, and gain structure in audio engineering.

dB(A) — A sound level measurement using the A-weighting curve to approximate human hearing sensitivity. It is widely used in environmental and occupational noise assessments. dB(A) emphasizes mid-frequencies and reduces low-frequency influence. It provides a standardized measure of perceived loudness.

dB(B) — A sound level measurement using the B-weighting curve, which approximates human hearing at moderate levels. Although less common today, dB(B) was historically used for certain noise assessments. It provides a middle ground between A- and C-weighting. Modern practice rarely employs B-weighting.

dB(C) — A sound level measurement using the C-weighting curve, which provides a relatively flat response across most of the audible spectrum. It is used for evaluating high-level or broadband noise where low-frequency content is significant. dB(C) measurements are common in industrial, environmental, and impact-noise assessments. The metric complements A-weighting by offering a more neutral frequency balance.

dBm — A unit expressing power level in decibels relative to one milliwatt. It is widely used in telecommunications, RF systems, and some audio equipment specifications. dBm provides an absolute reference for comparing signal strength. Its use ensures consistent interpretation across devices and applications.

dBu — A voltage-based decibel unit referenced to 0.775 volts RMS. It is commonly used to specify professional audio signal levels. dBu provides a standardized way to compare equipment performance and headroom. It is independent of load impedance, making it broadly applicable.

dBV — A decibel unit that expresses voltage level relative to **1 volt RMS**. Because it uses an absolute reference, dBV allows direct comparison of signal levels across devices without needing to know circuit impedance. A value of 0 dBV corresponds to exactly 1 V, while negative values indicate voltages below this reference. dBV is common in consumer and some professional audio specifications, helping engineers evaluate headroom, noise performance, and gain structure.

DC (Direct Current) — An electrical signal or power source in which voltage and current flow in a single, unchanging direction. In audio systems, DC is not part of the audible signal but appears in power supplies, bias circuits, and control voltages. Unwanted DC in an audio path can cause clicks, distortion, or loudspeaker damage, which is why many devices use capacitors or transformers for DC blocking. Understanding DC behavior is essential for protecting equipment and maintaining clean, stable signal paths.

Dead Room — A room with extremely high absorption and minimal reverberation. Dead rooms eliminate reflections, creating an acoustically "dry" environment. They are used for close-miked recording, speech testing, and certain types of analysis. Excessive deadness can feel unnatural for general listening.

Decade — A tenfold increase or decrease in frequency. Decades are used in filter design, measurement scaling, and acoustic modeling. They help describe broad spectral trends and system behavior. Logarithmic frequency analysis often uses decade spacing.

Decay — The reduction in sound level after a source stops emitting. Decay characterizes how quickly energy dissipates in a room or system. Smooth, controlled decay supports clarity and naturalness. Irregular decay can cause coloration or muddiness.

Decay Rate — The speed at which sound level decreases during the decay process, typically expressed in decibels per second. It helps quantify reverberation characteristics and room behavior. Faster decay improves intelligibility, while slower decay enhances spaciousness. Designers adjust materials and geometry to control decay rate.

Decibel — A logarithmic unit used to express ratios of sound pressure, power, or voltage. The decibel compresses large dynamic ranges into manageable numbers. It is fundamental to all acoustic and audio measurements. Interpretation depends on reference values and weighting curves.

Decibel Per Second — A measure of how quickly sound level changes over time. It is used in analyzing transient behavior, decay characteristics, and dynamic events. High rates indicate rapid changes such as impacts or bursts. The metric supports detailed acoustic analysis.

Deck — A device used for recording or playing back audio, traditionally referring to tape decks. Decks include mechanisms for transport, heads, and electronics. They were central to analog recording workflows. Modern usage may refer to digital playback or DJ equipment.

De-Emphasis — A filtering process that reduces high-frequency content to complement earlier pre-emphasis applied during recording or transmission. De-emphasis restores natural tonal balance and reduces noise. It is used in FM broadcasting, tape systems, and some digital formats. Proper matching ensures accurate reproduction.

De-Esser — A processor that reduces excessive sibilance in vocal recordings. It targets high-frequency bursts caused by "s," "sh," and similar sounds. De-essers improve clarity and listener comfort without dulling the overall tone. They are essential tools in vocal production.

Degausser — A device that erases magnetic recordings by applying a strong alternating magnetic field. Degausser use ensures tapes can be reused or securely erased. It also helps maintain equipment by demagnetizing tape heads. Proper degaussing prevents residual noise and distortion.

Delay — An audio effect or system function that stores a signal and plays it back after a set time. Delay creates echoes, spatial enhancement, and rhythmic effects. It is used in mixing, sound reinforcement, and acoustic alignment. Delay settings shape depth and timing relationships.

Delay Line — A device or algorithm that provides a controlled time delay for an audio signal. Delay lines support effects, synchronization, and system tuning. They may be analog, digital, or acoustic in nature. Their precision influences clarity and spatial imaging.

Demodulation — The process of extracting audio or data from a modulated carrier signal. Demodulation is essential in wireless microphones, radio systems, and communication networks. It restores the original signal for processing or playback. Proper demodulation ensures clarity and reliability.

Demodulator — A device or circuit that performs demodulation. It separates the audio or data content from the carrier wave. Demodulators vary by modulation type, such as AM, FM, or digital schemes. Their performance affects noise, distortion, and signal integrity.

Detent — A mechanical or tactile notch in a control knob or fader that indicates a specific position, such as center or unity gain. Detents help operators make precise adjustments without looking. They improve repeatability and workflow efficiency. Many mixers and processors use detents for key settings.

Deviation — The amount by which a signal varies from a reference value. In FM systems, deviation refers to frequency shift around the carrier. In measurement, it may describe error or variance. Understanding deviation helps maintain accuracy and compliance.

Diaphragm — A thin, flexible membrane that vibrates in response to sound or electrical signals. In microphones, diaphragms convert sound into electrical energy; in loudspeakers, they convert electrical energy into sound. Material and design influence sensitivity, frequency response, and distortion. The diaphragm is central to transducer performance.

Dielectric — An insulating material that separates conductive elements in capacitors and cables. Dielectric properties affect capacitance, signal integrity, and high-frequency behavior. High-quality dielectrics reduce loss and distortion. They are essential in precision audio components.

Diffracted Wave — A sound wave that bends around obstacles or spreads after passing through openings. Diffraction allows sound to reach areas not in direct line of sight. It becomes more pronounced at longer wavelengths. Understanding diffraction helps predict coverage and shadow zones.

Diffraction — The bending and spreading of sound waves when they encounter edges, openings, or obstacles. Diffraction influences how sound propagates in rooms, outdoors, and around structures. It is most significant at low frequencies. Designers account for diffraction in loudspeaker placement and architectural acoustics.

Diffuse Field — A sound field in which energy is uniformly distributed and arrives equally from all directions. Diffuse fields occur in highly reverberant or well-scattered environments. They support consistent measurement and analysis. Achieving a diffuse field is important in reverberation chambers.

Diffuse Sound Field — A practical realization of a diffuse field in a room or enclosure. It features random incidence of reflections and uniform energy distribution. Diffuse fields help evaluate absorption and scattering properties. They provide stable conditions for acoustic testing.

Diffusion — The scattering of sound waves to reduce strong reflections and improve spatial uniformity. Diffusion enhances clarity, envelopment, and naturalness in rooms. It is achieved through specialized surfaces or architectural features. Proper diffusion complements absorption for balanced acoustics.

Diffusion Coefficient — A measure of how effectively a surface scatters sound energy. Higher coefficients indicate stronger diffusion and reduced specular reflections. The value depends on surface geometry, frequency, and material. It guides the design of diffusers and acoustic treatments.

Diffusor — A device designed to scatter sound waves in many directions. Diffusors reduce echoes, standing waves, and coloration while preserving liveliness. They are used in studios, auditoriums, and listening rooms. Their shape and depth determine performance across frequencies.

Digital — Relating to signals or systems that represent information using discrete numerical values. Digital audio offers precision, flexibility, and resistance to noise. It enables advanced processing, storage, and transmission. Digital technology underpins modern audio production and distribution.

Digital Audio Tape — A digital recording format that stores audio on magnetic tape using helical-scan technology. DAT provided high-quality, portable digital recording in the 1990s. It supported 16-bit, 48 kHz audio and became a studio standard. Although largely obsolete, DAT remains important in archival work.

Digital Delay — A delay effect implemented using digital signal processing. It offers precise timing, long delay times, and flexible modulation options. Digital delay is used for echoes, spatial enhancement, and rhythmic effects. Its clarity and control surpass analog delay methods.

Digital Reverb — A reverberation effect created through digital algorithms that simulate reflections and decay patterns. Digital reverb models real spaces or creates entirely new acoustic environments. It is essential in music production, film, and sound design. Its flexibility allows detailed control over spatial character.

Digitization — The process of converting analog audio into digital form through sampling and quantization. Digitization enables editing, storage, and transmission in digital systems. Quality depends on sampling rate, bit depth, and converter design. It is the foundation of modern audio workflows.

Dip — A reduction or notch in frequency response caused by destructive interference, resonance issues, or filtering. Dips can make audio sound thin or uneven, especially if they occur in critical mid-frequency regions. They often result from room acoustics, microphone placement, or loudspeaker interactions. Identifying and correcting dips improves tonal balance and clarity.

Dipole — A sound source or microphone pattern with two opposite lobes of equal strength and opposite polarity. Dipoles radiate or capture sound from the front and rear while rejecting the sides. They exhibit strong directional behavior and unique low-frequency characteristics. Dipole loudspeakers and microphones are used for spatial realism and controlled dispersion.

Direct — Referring to sound or signal that travels straight from the source to the listener or device without reflections or processing. Direct sound provides clarity, localization cues, and intelligibility. In audio systems, maintaining a strong direct component improves accuracy. It contrasts with reverberant or reflected sound.

Direct Box — A device that converts high-impedance, unbalanced instrument signals into low-impedance, balanced signals suitable for mixers. Direct boxes reduce noise, allow long cable runs, and provide proper impedance matching. They may include ground lifts and attenuation pads. DI boxes are essential in live sound and recording.

Direct Damage to Major Hearing Structures — Injury to the peripheral auditory system caused by intense mechanical, acoustic, or chemical forces that exceed the tissue's capacity to recover. This form of trauma can disrupt the tympanic membrane, dislocate or fracture the ossicles, tear the oval or round windows, or destroy hair cells within the cochlea. Severe cases may also compromise the stria vascularis, spiral ganglion neurons, or vestibular organs, leading to permanent threshold elevation, distortion, or loss of frequency selectivity. Unlike temporary metabolic fatigue, this category of injury reflects structural failure, and its consequences are typically irreversible without surgical intervention.

Direct Field — The region close to a sound source where direct sound dominates over reflections. In the direct field, sound level decreases predictably with distance. This region provides the clearest representation of the source. Microphone placement often targets the direct field for accuracy.

Direct Output — A dedicated output on a mixer or device that sends an unprocessed or lightly processed signal from a specific channel. Direct outputs support multitrack recording, routing flexibility, and signal isolation. They preserve channel integrity for later mixing. Their configuration varies by console design.

Direct Sound — The portion of sound that reaches the listener without reflecting off surfaces. Direct sound carries the most accurate timing, localization, and tonal information. It is essential for speech intelligibility and musical clarity. Its strength relative to reverberation shapes perceived quality.

Direct Sound Field — A sound field dominated by direct sound rather than reflections or reverberation. It exists close to the source or in highly absorptive environments. The direct sound field supports precise measurement and critical listening. Its boundary is defined by the critical distance.

Directivity Factor — A measure of how much a sound source concentrates energy in a particular direction compared to an omnidirectional source. Higher directivity factors indicate more focused radiation. This metric helps predict coverage, gain-before-feedback, and room interaction. It is fundamental in loudspeaker design.

Directivity Index — The logarithmic expression of directivity factor, given in decibels. It quantifies how directional a loudspeaker or microphone is. Higher DI values indicate narrower coverage and greater focus. The index guides system tuning and placement.

Direct-to-Reverberant Ratio — The ratio of direct sound energy to reverberant energy at a listening position. A high ratio improves clarity, intelligibility, and localization. A low ratio results in muddiness or loss of detail. The metric is central to room acoustics and sound reinforcement.

Discrete Tone — A single, well-defined frequency component within a sound. Discrete tones stand out from broadband noise and are easily identifiable. They are used in testing, calibration, and psychoacoustic research. Excessive discrete tones in noise environments may be regulated due to annoyance.

Dispersion — The spread of sound energy across angles or frequencies. Loudspeaker dispersion determines coverage uniformity and tonal consistency. Controlled dispersion improves clarity and reduces unwanted reflections. Designers shape dispersion through horns, waveguides, and array geometry.

Displacement — The physical movement of particles or surfaces in response to sound waves. In loudspeakers, diaphragm displacement determines low-frequency output and power handling. Excessive displacement can cause distortion or mechanical failure. Displacement is a key parameter in acoustic modeling.

Distance Double Law — A principle stating that sound pressure level decreases by 6 dB each time the distance from a point source doubles in a free field. This inverse-square behavior applies only when reflections are minimal. It helps predict level changes and system coverage. Deviations occur in reverberant or complex environments.

Distant Pickup — A microphone technique where the mic is placed far from the sound source to capture room ambience and natural balance. Distant pickup emphasizes reverberation and spatial cues. It is common in orchestral, choral, and environmental recording. Proper placement avoids excessive noise or loss of clarity.

Distortion — Any alteration of an audio signal that changes its waveform from the original. Distortion may arise from clipping, nonlinearities, or mechanical limitations. Some forms are undesirable, while others are used creatively. Minimizing distortion preserves fidelity and transparency.

Distortion Harmonic — Harmonic frequencies generated by nonlinear behavior in audio equipment or transducers. These harmonics add coloration and reduce accuracy. Harmonic distortion is measured to evaluate equipment quality. Lower levels indicate cleaner reproduction.

Distribution Amplifier — A device that splits an audio signal into multiple outputs while maintaining level and quality. Distribution amplifiers prevent loading effects and signal degradation. They are used in broadcast, installed sound, and measurement systems. Proper design ensures consistent performance across outputs.

Divergence — The spreading of sound energy as it propagates from a source. Divergence reduces sound intensity with distance and influences coverage patterns. It is governed by geometry and environment. Understanding divergence helps predict level distribution.

Diversity — A technique in wireless audio systems that uses multiple antennas or receivers to reduce dropouts. Diversity improves reliability by selecting the strongest signal path. It mitigates interference, reflections, and fading. Professional wireless systems rely on diversity for stable performance.

Dividing Network — Another term for a crossover network that splits audio into frequency bands for different loudspeaker drivers. Dividing networks ensure each driver operates within its optimal range. They may be passive or active. Proper design improves efficiency and tonal balance.

Dolby Digital — A digital audio encoding format that supports multichannel surround sound. It compresses audio efficiently while preserving clarity and spatial detail. Dolby Digital is widely used in film, broadcast, and home theater. It supports formats up to 5.1 channels.

Dolby Prologic — A matrix-based surround sound system that derives multichannel audio from stereo sources. Pro Logic decodes directional cues to create center, surround, and sometimes subwoofer channels. It enhances spatial immersion without requiring discrete multichannel content. It was a major step in home theater development.

Doppler — Referring to the change in frequency of a sound due to relative motion between source and listener. The Doppler effect causes rising pitch as a source approaches and falling pitch as it recedes. It is fundamental in acoustics, radar, and astronomy. Doppler shifts reveal motion and velocity.

Doppler Effect — The perceived change in frequency caused by motion of the source, listener, or medium. It explains pitch changes in passing vehicles, sirens, and moving sound sources. The effect is used in medical imaging, navigation, and scientific measurement. It illustrates the relationship between motion and wave behavior.

Doubler — An audio effect or device that creates the impression of two simultaneous performances by duplicating and slightly delaying or modulating a signal. Doublers add thickness, width, and richness to vocals and instruments. They simulate natural variations between performers. The effect is common in music production.

Drain Wire — A conductor in shielded cable that provides a low-impedance path to ground. Drain wires simplify termination and improve shielding effectiveness. They help reduce hum, RF interference, and noise. Proper grounding ensures clean audio transmission.

Driver — A loudspeaker component that converts electrical signals into sound. Drivers include woofers, midrange units, and tweeters, each optimized for specific frequency ranges. Their design affects efficiency, distortion, and tonal character. Drivers are the core elements of loudspeaker systems.

Drop Out — A brief loss of audio signal caused by interference, equipment failure, or media defects. Dropouts disrupt continuity and reduce reliability. They are common in wireless systems, tape formats, and digital transmission. Proper system design minimizes dropout risk.

Dry Signal — An audio signal without added effects such as reverb, delay, or modulation. Dry signals provide clarity and serve as the foundation for further processing. Engineers blend dry and wet signals to shape spatial and tonal character. Dry recording preserves flexibility in mixing.

DSP — A field of engineering that analyzes, transforms, and synthesizes signals using mathematical operations implemented in software or dedicated hardware. It enables filtering, equalization, compression, spatialization, and countless other audio processes by manipulating discrete-time samples rather than continuous waveforms. Modern systems rely on fast algorithms such as the FFT, efficient filter structures, and optimized numerical formats to achieve real-time performance. Its flexibility and precision make it foundational to contemporary acoustics, recording, telecommunications, and embedded audio devices.

DTS — A family of multichannel audio encoding technologies developed to deliver high-fidelity surround sound for cinema, home theater, gaming, and streaming applications. It uses higher bitrates and flexible compression strategies to preserve spatial detail, dynamic range, and clarity across multiple playback environments. Variants such as DTS-HD Master Audio and DTS:X extend the core format with lossless delivery and object-based rendering, enabling precise three-dimensional placement of sound sources. Its ecosystem spans theatrical distribution, Blu-ray authoring, consumer electronics, and immersive audio production workflows.

Dub — A process of copying or transferring audio from one medium or track to another. Dubbing is used in film, music production, and archival work. It may also refer to a musical style characterized by heavy effects and remixing. Proper dubbing preserves quality and synchronization.

Duct Lining — A sound-absorbing material installed inside HVAC ducts to reduce airflow noise and vibration. It helps control fan noise, turbulence, and mechanical transmission through ductwork. Proper lining improves acoustic comfort in buildings without restricting airflow. Its performance depends on thickness, density, and installation quality.

DVD — An optical disc format designed to store digital video, audio, and data at significantly higher capacity than earlier CD-based media. It uses tighter laser focusing, multi-layer encoding, and more efficient error-correction schemes to deliver full-resolution video and multichannel audio for home entertainment and professional distribution. Variants such as DVD-Video and DVD-Audio support specialized authoring structures, menu systems, and high-fidelity playback. Although largely superseded by Blu-ray and streaming platforms, it remains widely used for archival content, educational materials, and legacy playback systems.

Dynamic Equalization — An EQ process that changes frequency response based on signal level. It boosts or cuts frequencies only when needed, preserving natural tone during quieter passages. Dynamic EQ helps control harshness, muddiness, or resonance without overprocessing. It is widely used in mixing and mastering for transparent correction.

Dynamic Headroom — The amount of extra level an audio system can handle above its nominal operating point without distortion. Greater headroom allows clean reproduction of peaks and transients. It is essential for high-quality recording, amplification, and playback. Systems with limited headroom clip more easily.

Dynamic Microphone — A microphone that uses electromagnetic induction to convert sound into electrical signals. Dynamic mics are durable, handle high sound pressure levels, and resist moisture. They are ideal for live sound, drums, and amplified instruments. Their sound is typically robust and less sensitive than condenser microphones.

Dynamic Range — The difference between the quietest and loudest sounds a system can reproduce without noise or distortion. A wide dynamic range allows detailed, expressive audio. It is influenced by noise floor, headroom, and equipment quality. Dynamic range is critical in recording, mixing, and playback.

Dynamics — The variations in loudness within a performance or audio signal. Dynamics shape musical expression, emotional impact, and clarity. Engineers manage dynamics through compression, limiting, and automation. Balanced dynamics enhance intelligibility and listener engagement.

Dyne — A unit of force in the centimeter-gram-second (CGS) system. In acoustics, it appears in older literature describing pressure or force relationships. One dyne equals 10^{-5} newtons. Modern practice uses SI units, but the dyne remains historically relevant.

Ear Canal — The passage leading from the outer ear to the eardrum. It amplifies certain frequencies and helps direct sound into the middle ear. Its shape influences resonance and hearing sensitivity. Protecting the ear canal is essential for hearing health.

Eardrum — A thin membrane that vibrates in response to sound waves entering the ear canal. It transmits these vibrations to the ossicles in the middle ear. The eardrum is crucial for converting airborne sound into mechanical motion. Damage to it can impair hearing.

Earmuff — A hearing protection device that covers the entire outer ear to reduce noise exposure. Earmuffs use padded cups and insulation to block sound. They are effective in industrial, recreational, and aviation environments. Proper fit ensures maximum protection.

Earphone — A small audio transducer worn in or on the ear for personal listening. Earphones vary in design, including earbuds and in-ear monitors. Their performance depends on fit, driver quality, and isolation. They are widely used for music, communication, and monitoring.

Earplug — A small insert worn in the ear canal to reduce sound exposure. Earplugs provide effective protection in loud environments and help prevent hearing damage. They may be disposable, reusable, or custom-molded. Proper insertion is essential for full attenuation.

Echo — A distinct reflection of sound that arrives after a noticeable delay from the direct sound. Echoes occur in large or reflective spaces where sound travels long distances. They can enhance spatial impression or cause intelligibility problems. Controlling echo is important in architectural acoustics.

Echo Location — A biological process in which animals emit sounds and interpret returning echoes to navigate or locate objects. Bats, dolphins, and some birds rely heavily on echolocation. The process reveals distance, size, and movement of targets. It demonstrates advanced natural acoustic sensing.

Echograms — Visual representations of echo patterns over time, often used in sonar, ultrasound, or room acoustics. Echograms show reflection timing, strength, and structure. They help analyze environments, materials, and spatial characteristics. Engineers use them to diagnose acoustic issues.

EES — A term used in architectural and environmental acoustics to describe *Early, Early Sound*, referring to the first cluster of reflections that arrive immediately after the direct sound but before the conventional early-reflection window. These micro-reflections shape clarity, presence, and spatial impression by reinforcing or subtly coloring the direct signal. Their timing and spectral content influence speech intelligibility, musical articulation, and listener envelopment, especially in small to mid-sized rooms. Designers analyze this region to optimize surface geometry, diffusion, and material placement for precise acoustic imaging.

EFC — A representation of how acoustic energy varies across frequency, typically used to analyze the spectral balance, efficiency, or resonant behavior of a system, room, or transducer. It highlights which frequency bands dominate, which are suppressed, and how energy distribution shifts under different operating conditions. Engineers use this curve to diagnose coloration, identify modal issues, and evaluate the performance of loudspeakers, microphones, and architectural spaces. Its shape provides insight into both tonal character and underlying physical mechanisms, making it a foundational tool in acoustic measurement and system tuning.

Effective Sound Pressure — A root-mean-square (RMS) measure of sound pressure that represents the average energy of a sound wave. It provides a stable basis for calculating sound pressure level. Effective sound pressure smooths rapid fluctuations in waveform amplitude. It is fundamental to acoustic measurement.

Effects — Audio processes that modify sound through reverb, delay, modulation, distortion, or other transformations. Effects shape tone, space, and character in music and sound design. They may be applied subtly or dramatically depending on artistic intent. Effects are central to modern production workflows.

Effects Return — A mixer input dedicated to receiving processed signals from external effects units. It allows engineers to blend wet signals with the original dry audio. Effects returns often include level and routing controls. They streamline integration of outboard processing.

Efficiency — A measure of how effectively a loudspeaker converts electrical power into acoustic output. Higher efficiency means more sound for the same input power. Efficiency influences system design, amplifier requirements, and overall performance. It is especially important in large-scale sound reinforcement.

EFTC — A metric used in room and architectural acoustics to describe the *Energy–Frequency Time Curve*, a representation that tracks how acoustic energy decays across frequency bands as a function of time. It combines spectral and temporal information, allowing engineers to see not only which frequencies dominate but also how quickly each band dissipates within a space. This dual-axis perspective helps identify modal buildup, flutter regions, and uneven decay patterns that affect clarity, warmth, and intelligibility. Because it reveals both spectral coloration and temporal smear, it is often used in conjunction with reverberation-time plots and waterfall analyses to guide treatment strategies.

EFX — A shorthand label commonly used in audio production and acoustics to denote *effects processing*, referring to any signal-chain stage that alters timbre, space, dynamics, or spectral content. It encompasses tools such as reverbs, delays, modulation units, distortion circuits, and spatial processors that shape the artistic or functional character of a sound. In technical workflows, the term often appears in routing diagrams, console labeling and DSP block structures to distinguish dry signal paths from processed ones. Because it spans both creative and corrective domains, it serves as a flexible umbrella for any transformation applied after the original source is captured or generated.

EIN — A measure of the inherent noise produced by an electronic audio device, expressed as the equivalent input noise that would generate the same output level if applied at the input. It allows engineers to compare the self-noise performance of preamps, microphones, and other front-end circuitry independent of gain settings. Lower values indicate cleaner, quieter operation, especially important when working with low-output transducers or high-dynamic-range material. Because it reflects both circuit topology and component quality, it serves as a key benchmark for evaluating professional recording and measurement equipment.

Electret-Condensor Microphone — A condenser microphone that uses a permanently charged electret material instead of requiring external bias voltage. Electret mics are compact, inexpensive, and widely used in consumer devices. They offer good sensitivity and frequency response for their size. Their convenience makes them popular in mobile and embedded applications.

Electroacoustic Transducer — A device that converts electrical energy into sound or vice versa. Examples include microphones, loudspeakers, and headphones. Transducer design determines sensitivity, frequency response, and distortion. They form the core of all audio capture and reproduction systems.

Energy Average Sound Level — A long-term average sound level calculated using energy-based (RMS) averaging. It reflects cumulative exposure rather than instantaneous peaks. This metric is used in environmental noise assessment and occupational safety. It provides a realistic measure of sustained sound levels.

Energy Decay Curve — A graph showing how sound energy decreases over time after a source stops. It is used to calculate reverberation time and analyze room acoustics. Smooth decay indicates good acoustic behavior, while irregular decay reveals problems. The curve is central to room measurement techniques.

Ensemble — A group of performers or sound sources working together in a coordinated musical or acoustic context. Ensemble performance relies on balance, timing, and spatial interaction. Acoustics influence how ensembles blend and project. Recording ensembles requires careful microphone placement and room design.

EQ — A process that adjusts the balance of frequency components within an audio signal to shape clarity, tone, and overall spectral character. It can correct problems such as muddiness, harshness, or masking, or it can be used creatively to emphasize desirable qualities in voices, instruments, and environments. Implementations range from simple tone controls to complex parametric, graphic, and linear-phase designs, each offering different levels of precision and transparency. Because it influences both technical accuracy and artistic expression, it is one of the most fundamental tools in acoustics, recording, and sound reinforcement.

Equal Loudness Contour — A curve showing the sound pressure levels required at different frequencies for a listener to perceive equal loudness. These contours reveal the ear's varying sensitivity across the spectrum. They guide weighting filters, audio design, and psychoacoustic modeling. The most common set is the ISO 226 standard.

Equalization — The process of adjusting frequency content to shape tonal balance or correct acoustic issues. Equalization can enhance clarity, reduce resonances, or tailor sound for artistic effect. It is applied through filters, parametric EQs, and graphic EQs. EQ is essential in mixing, mastering, and system tuning.

Equalizer — A device or software tool that provides control over specific frequency bands. Equalizers allow precise shaping of tonal character and correction of acoustic problems. They come in graphic, parametric, and shelving designs. Equalizers are indispensable in both live and studio environments.

Equivalent Continuous A-Weighted Sound Pressure Level — A time-averaged sound level using A-weighting to reflect human hearing sensitivity. It represents cumulative noise exposure over a defined period. This metric is widely used in environmental and occupational noise regulation. It provides a realistic measure of perceived loudness over time.

Equivalent Sound Level — A long-term average sound level that represents fluctuating noise as a single steady value. It is calculated using energy averaging and may include weighting filters. Equivalent sound level is central to environmental noise assessment. It simplifies comparison of complex noise environments.

ETC — The Energy Time Curve, a graph that shows how sound energy arrives over time at a measurement position. It highlights the timing and strength of early reflections relative to the direct sound, making it a powerful tool for analyzing clarity, imaging, and acoustic defects. Engineers use the ETC to identify problematic reflections, optimize loudspeaker placement, and evaluate room treatment. A well-interpreted ETC helps improve intelligibility, spatial accuracy, and overall acoustic performance.

Euphonic — Describing sound that is pleasing, warm, or musically satisfying to the ear. Euphonic qualities often arise from subtle harmonic coloration or gentle nonlinearities in equipment. Some listeners prefer euphonic reproduction over strictly neutral sound. The term reflects subjective perception rather than strict accuracy.

Eustachian Tube — A narrow passage connecting the middle ear to the back of the throat. It equalizes pressure on both sides of the eardrum, ensuring proper vibration and hearing function. Blockage can cause discomfort, reduced hearing, or infections. Its operation is essential for maintaining auditory balance.

Expansion — A dynamic process that increases the difference between quiet and loud parts of a signal. Expansion reduces low-level noise and enhances clarity by lowering the volume of soft sounds relative to louder ones. It is used in noise gates, broadcast processing, and restoration work. Proper settings prevent unnatural pumping or artifacts.

Extension — The ability of an audio system to reproduce very low or very high frequencies beyond the typical range. Good low-frequency extension adds depth and power, while high-frequency extension adds air and detail. Extension depends on driver design, enclosure tuning, and system bandwidth. It is a key factor in perceived fidelity.

External Meatus — Another term for the external auditory canal, the passage leading from the outer ear to the eardrum. It shapes incoming sound through resonance and directional filtering. Its geometry influences hearing sensitivity and localization cues. Protecting the external meatus is important for hearing health.

Eyring Equation — A formula used to estimate reverberation time in rooms with moderate to high absorption. It improves on the Sabine equation by accounting for non-uniform absorption and higher loss factors. The Eyring model provides more accurate predictions in many real-world spaces. It is widely used in architectural acoustics.

Fade — A gradual increase or decrease in audio level over time. Fades smooth transitions between sounds, prevent abrupt endings, and shape emotional impact. They are essential in mixing, broadcasting, and live performance. Well-executed fades enhance flow and listener comfort.

Fader — A sliding control on a mixer or interface used to adjust audio level. Faders provide precise, intuitive control over gain and balance. Their travel range allows smooth transitions and fine adjustments. They are central to mixing workflows.

Fan — A mechanical device used to move air for cooling electronic equipment. Fans can introduce noise into audio environments if not properly isolated. Quiet, low-vibration designs are preferred in studios and control rooms. Managing fan noise is part of acoustic system design.

Far Field — The region where sound waves have spread sufficiently that their behavior approximates plane waves. In the far field, sound level decreases predictably with distance, and directivity patterns stabilize. Measurements taken in the far field reflect true source characteristics. It contrasts with the near field, where complex interactions occur.

FCC — The independent U.S. regulatory agency responsible for overseeing interstate and international communications across radio, television, wire, satellite, and broadband services. It allocates spectrum, enforces technical and operational standards, and manages licensing to ensure reliable, interference-free communication systems. Its policies shape everything from broadcast loudness rules to wireless-microphone allocations and consumer-device compliance. Because its decisions directly affect audio, media, and telecommunications industries, it remains a central authority in defining technical, legal, and operational frameworks for modern communication infrastructure.

Feed — A signal sent from one device or system to another. Feeds may include program audio, monitor mixes, or communication channels. Proper routing ensures clarity and prevents feedback or overload. Feeds are fundamental to audio distribution.

Feedback — A loud, often unpleasant tone caused when a microphone picks up sound from a loudspeaker and re-amplifies it in a loop. Feedback occurs when system gain exceeds stable limits. It is influenced by microphone placement, room acoustics, and loudspeaker directivity. Managing feedback is essential in live sound.

Feedback Acoustic — A term describing feedback specifically caused by acoustic coupling between microphones and loudspeakers. It highlights the role of room reflections and direct sound paths. Acoustic feedback can be mitigated through EQ, placement, and directional control. It is a major concern in reinforcement systems.

Feedback Eliminator — A device or processor that detects and suppresses feedback frequencies automatically. It uses notch filters or adaptive algorithms to reduce problem peaks. Feedback eliminators improve stability in challenging acoustic environments. They are common in live sound and conferencing systems.

FFT — A computational algorithm that rapidly computes the Discrete Fourier Transform, converting a time-domain signal into its frequency-domain representation with far greater efficiency than direct DFT calculation. Its speed enables real-time spectral analysis, filtering, convolution, and system identification across audio, acoustics, and DSP applications. By breaking the transform into smaller, recursively structured operations, it reduces computational complexity from N^2 to $N\log N$, making large-scale processing practical on modern hardware. Because it underpins spectrograms, equalization, room-analysis tools, and countless DSP blocks, it remains one of the most foundational algorithms in signal processing.

Fiber Glass Insulation — A porous material made of fine glass fibers used for thermal and acoustic insulation. It absorbs sound effectively across mid- and high-frequencies. Fiberglass is widely used in walls, ceilings, and acoustic panels. Proper installation ensures safety and performance.

Fiber Optic — A transmission medium that uses light to carry digital audio or data signals. Fiber optic cables offer immunity to electromagnetic interference and support long distances. They are used in high-end audio networks, broadcast, and telecommunications. Fiber ensures reliable, high-bandwidth signal transport.

Fidelity — The degree to which an audio system accurately reproduces the original sound. High fidelity preserves detail, dynamics, and tonal balance. Fidelity depends on equipment quality, acoustics, and signal integrity. It is a central goal in audio engineering.

Field Sound Transmission Class — A rating that measures the sound insulation performance of building elements in real-world conditions. Unlike laboratory STC, it accounts for flanking paths and construction variability. Field STC provides a realistic assessment of acoustic privacy. It is used in building evaluation and compliance.

Field Transmission Loss — The measured reduction in sound level as it passes through a building element in actual field conditions. It reflects real-world installation quality and flanking effects. Field TL values guide acoustic design and troubleshooting. They often differ from laboratory results.

Filter — A device or algorithm that modifies frequency content by allowing certain frequencies to pass while attenuating others. Filters shape tone, remove noise, and support system tuning. Types include low-pass, high-pass, bandpass, and notch filters. Filtering is fundamental to audio processing.

Filter Bandpass — A filter that passes a specific range of frequencies while attenuating those above and below it. Bandpass filters isolate tonal regions or remove unwanted content. They are used in measurement, effects, and communication systems. Their bandwidth and center frequency define performance.

Filter High Pass — A filter that allows frequencies above a cutoff point to pass while reducing lower frequencies. High-pass filters remove rumble, plosives, and low-frequency noise. They are essential in mixing, live sound, and measurement. Proper use improves clarity and headroom.

Filter Low Pass — A filter that allows frequencies below a cutoff point to pass while attenuating higher frequencies. Low-pass filters shape tone, remove hiss, and support crossover design. They are used in synthesis, mixing, and system tuning. Cutoff frequency determines tonal character.

Filter Nominal Midband Frequency — The central frequency around which a filter's passband is defined. It represents the midpoint of effective transmission. This value helps characterize bandpass and parametric filters. Accurate midband specification ensures predictable behavior.

Finite Difference Time Domain — A numerical method that simulates sound propagation by solving wave equations over discrete time and space steps. FDTD models complex acoustic environments with high accuracy. It is used in research, product design, and architectural acoustics. The method requires significant computational resources.

Finite Element Method — A computational technique that divides structures or spaces into small elements to solve acoustic or structural equations. FEM handles complex geometries and material properties. It is widely used in transducer design, room acoustics, and vibration analysis. FEM provides detailed insight into acoustic behavior.

Fishpole — A long boom pole used to position microphones in film, broadcast, and field recording. Fishpoles allow close placement while keeping equipment out of frame. They require skill to minimize handling noise and maintain stability. Lightweight materials improve comfort and control.

Fixed EQ — A tone-shaping circuit or digital filter whose frequency centers, bandwidths, and gain steps are predetermined rather than user-selectable. It provides consistent, repeatable spectral adjustments optimized for specific applications such as broadcast chains, instrument preamps, or consumer playback devices. Because its parameters cannot be tuned, it emphasizes reliability, speed, and voicing over surgical precision. This design is often chosen when predictable tonal character, minimal user intervention, or hardware simplicity is more important than detailed frequency control.

Flame Spread — A measure of how quickly fire travels across a material's surface. In acoustics, flame-spread ratings determine the safety of acoustic panels and treatments. Building codes require materials to meet specific flame-spread classifications. Safe materials reduce fire risk in public spaces.

Flanging — An audio effect created by mixing a signal with a delayed, modulated copy of itself. Flanging produces sweeping comb-filter patterns and a distinctive swirling sound. It is used creatively in music production for movement and texture. The effect originated from manipulating tape machines.

Flanking — The transmission of sound through indirect paths such as walls, floors, or structural elements. Flanking bypasses primary barriers and reduces overall sound isolation. It is a major concern in building acoustics. Proper design and sealing minimize flanking paths.

Flanking Sound Transmission — The portion of sound that travels through secondary or structural paths rather than directly through a partition. It can significantly degrade isolation performance. Identifying flanking paths is essential for accurate acoustic diagnosis. Solutions involve structural decoupling and improved construction.

Flanking Transmission — A general term describing sound transmission through unintended or indirect building paths. It affects walls, ceilings, floors, and mechanical systems. Flanking transmission often explains discrepancies between laboratory and field performance. Effective mitigation requires holistic building design.

Flat — Describing an audio system or response that reproduces all frequencies at equal level without boosts or cuts. A flat response is valued for accuracy because it preserves the original tonal balance of the source. Engineers use flat monitoring to make reliable mixing decisions. True flatness is difficult to achieve due to room acoustics and equipment limitations.

Fletchermunson Curve — A set of equal-loudness contours showing how human hearing sensitivity varies with frequency and sound level. These curves reveal that the ear is less sensitive to very low and very high frequencies at low listening levels. They guide weighting filters, loudness compensation, and psychoacoustic modeling. Modern standards refine the original Fletcher-Munson data but preserve its core insights.

Floor Pocket — A recessed floor-mounted box that houses audio, power, or data connectors. Floor pockets provide clean, safe access to wiring in stages, auditoriums, and conference rooms. They reduce cable clutter and trip hazards. Proper installation ensures durability and signal integrity.

Flow Noise — Noise generated by the movement of air or fluid through ducts, vents, or mechanical systems. Turbulence, obstructions, and high velocities increase flow noise. It can interfere with speech intelligibility and acoustic comfort. Designers control flow noise through duct sizing, lining, and smooth airflow paths.

Flow-Induced Noise — Sound produced when airflow interacts with surfaces, edges, or objects. It includes whistling, rumbling, and turbulence-related noise. Flow-induced noise affects HVAC systems, vehicles, and outdoor environments. Reducing it improves comfort and system performance.

Fluctuating Noise — Noise whose level varies significantly over time. It may include intermittent, pulsating, or irregular patterns. Fluctuating noise is often more annoying than steady noise due to its unpredictability. Environmental assessments account for these variations using specialized metrics.

Fluctuation Strength — A psychoacoustic measure describing how strongly a listener perceives slow amplitude variations in a sound. It is most pronounced at modulation frequencies around 4 Hz. High fluctuation strength can make noise more intrusive or unpleasant. The metric supports sound quality evaluation and product design.

Flutter — A rapid, repetitive variation in sound level or pitch caused by mechanical instability or acoustic reflections. In tape machines, flutter refers to speed irregularities; in rooms, it relates to repeated reflections. Flutter degrades clarity and tonal stability. Proper design minimizes flutter effects.

Flutter Echo — A distinct, rapid series of echoes caused by sound bouncing between two parallel reflective surfaces. It produces a metallic or ringing quality that interferes with clarity. Flutter echo is common in untreated rooms with hard walls. Diffusion or absorption eliminates the effect.

Flying Fader — A motorized fader that moves automatically under digital control. Flying faders allow precise automation of level changes during mixing. They improve repeatability and workflow efficiency. They are standard in high-end consoles and digital workstations.

FM — A method of analog modulation in which the instantaneous frequency of a carrier wave varies according to the amplitude of the input signal. This approach provides strong resistance to noise and interference because information is encoded in frequency deviations rather than amplitude changes. It is widely used in broadcast radio, wireless communication, and certain audio-synthesis techniques where its smooth, wideband behavior produces characteristic tonal richness. Its performance depends on deviation ratio, bandwidth, and the linearity of both the modulator and demodulator stages.

Foil Shield — A thin metallic foil used in cables to block electromagnetic interference. Foil shields provide continuous coverage and excellent high-frequency protection. They are often combined with braided shields for improved performance. Proper shielding ensures clean, noise-free audio transmission.

Foldback — A monitor mix sent to performers on stage so they can hear themselves and other instruments. Foldback systems use wedges, in-ear monitors, or side fills. Good foldback improves timing, pitch accuracy, and performance confidence. Engineers balance foldback levels to avoid feedback.

Fourier Analysis — A mathematical method that decomposes complex signals into their frequency components. It reveals spectral content, harmonics, and periodicity. Fourier analysis underpins modern audio processing, measurement, and synthesis. It is essential for understanding how sound behaves in both time and frequency domains.

Free Field — An environment where sound propagates without reflections, as if in open space. In a free field, sound level decreases predictably with distance according to the inverse-square law. Free-field conditions are used for accurate acoustic measurements. Anechoic chambers approximate free-field behavior.

Free Progressive Wave — A sound wave that travels outward from a source without reflection or interference. It maintains consistent phase relationships and predictable decay. Free progressive waves occur in open environments or idealized models. They form the basis of many acoustic calculations.

Frequency — The number of cycles a sound wave completes per second, measured in hertz. Frequency determines pitch and influences how sound interacts with materials and spaces. Low frequencies produce bass, while high frequencies create brightness and detail. Frequency is a fundamental descriptor of all audio signals.

Frequency Response — A measure of how an audio system reproduces different frequencies relative to one another. A smooth, accurate response ensures natural tonal balance. Irregularities cause coloration or distortion. Frequency response is central to evaluating microphones, speakers, and electronics.

Frequency Response Curve — A graphical representation of a system's output level across the frequency spectrum. Peaks and dips reveal resonances, deficiencies, or design characteristics. Engineers use response curves to tune systems and diagnose issues. A flat curve indicates accurate reproduction.

Frequency Weighting — A filtering method that adjusts sound measurements to reflect human hearing sensitivity. Common weightings include A, C, and Z. Weighting ensures that measurements correspond to perceived loudness. It is essential in noise regulation and assessment.

Frequency-Dependent Directivity — The change in a sound source's directional behavior as frequency varies. Higher frequencies tend to beam more narrowly, while lower frequencies spread widely. Understanding this behavior helps optimize coverage and reduce reflections. It is crucial in loudspeaker design and placement.

Front End — The initial stage of an audio system where signals are captured or received. The front end includes microphones, preamps, and input circuitry. Its quality strongly influences noise, clarity, and overall fidelity. A strong front end ensures clean, accurate signal capture.

Frontcountry — A term describing accessible outdoor areas with moderate human activity and infrastructure. Frontcountry acoustics include both natural and human-made sounds. These environments are studied for recreation planning and noise management. They contrast with remote backcountry soundscapes.

FTC — A shorthand for *Frequency-Time Curve*, a representation that shows how the energy of an acoustic signal evolves across frequency bands over time. Unlike a simple frequency response, which is static, the FTC captures transient behavior, decay characteristics, and spectral shifts as a sound propagates through a room or system. Engineers use it to diagnose modal buildup, flutter, and uneven decay patterns that affect clarity, warmth, and intelligibility. Because it links spectral content with temporal dynamics, it serves as a bridge between traditional frequency-domain plots and full spectrogram or waterfall analyses.

Fuel Contributed — A fire-safety term describing the amount of combustible material a product adds to a fire. In acoustics, it applies to evaluating the safety of acoustic panels and treatments. Materials with low fuel contribution reduce fire risk. Building codes specify acceptable limits.

Full Range — Describing a loudspeaker or system capable of reproducing the entire audible spectrum. Full-range systems simplify setup but may sacrifice precision in certain bands. They are common in portable PA systems and compact designs. High-performance systems often supplement full-range units with subwoofers or tweeters.

Fundamental — The lowest frequency component of a periodic sound. The fundamental determines perceived pitch and anchors the harmonic structure. Higher harmonics shape timbre and character. Accurate reproduction of the fundamental is essential for natural sound.

Fundamental Frequency — The frequency corresponding to the fundamental component of a periodic waveform. It defines the base pitch of musical notes and speech sounds. Variations in fundamental frequency convey melody, intonation, and emotion. It is central to psychoacoustics and signal analysis.

Fuse — A safety device that protects electrical circuits by breaking the connection when current exceeds safe limits. Fuses prevent equipment damage and fire hazards. They must be properly rated for voltage and current. Replacing a blown fuse requires addressing the underlying cause.

Fusion Zone — The region where early reflections arrive quickly enough after the direct sound that the listener perceives them as a single fused event. Fusion enhances loudness and spatial impression without causing echo. The size of the fusion zone depends on timing and room geometry. It is important in sound reinforcement and acoustic design.

FX — A broad shorthand for *effects processing* applied to an audio signal to modify its spectral balance, spatial characteristics, dynamics, or timbral identity. In acoustics and sound-system design, FX paths include reverberation, delay, modulation, distortion, and spatial processors that shape perceived envelopment, clarity, and tonal color. FX routing often appears as inserts, aux sends, or dedicated processing blocks in consoles and DSP units, allowing engineers to blend dry and processed signals to achieve the desired acoustic impression.

Gain — The increase in signal level provided by an amplifier or processor. Proper gain staging ensures clean, noise-free audio with adequate headroom. Excessive gain causes distortion, while insufficient gain raises noise. Gain control is fundamental to all audio workflows.

Gain-Before-Feedback — The maximum system gain achievable before feedback occurs. Higher gain-before-feedback improves clarity and loudness in reinforcement systems. It depends on microphone placement, loudspeaker directivity, and room acoustics. Optimizing this parameter is essential for stable performance.

Galerkin Method — A numerical technique used to solve differential equations by approximating solutions with weighted basis functions. In acoustics, it supports modeling of wave propagation, vibration, and structural interaction. The method provides accurate results for complex geometries. It is widely used in computational acoustics.

Gammatone Filter — A filter model based on human auditory filtering characteristics. Gammatone filters simulate how the cochlea analyzes sound into frequency bands. They are used in speech processing, psychoacoustics, and auditory modeling. The filter bank reflects biological frequency resolution.

Gate — A dynamics processor that reduces or eliminates low-level noise by muting signals below a threshold. Gates clean up tracks by removing hiss, bleed, or room noise. Attack, release, and hold settings shape behavior. Gates are essential in live sound and studio production.

Gauge — A measurement standard used to specify the thickness, diameter, or capacity of materials such as wire or metal. In audio, gauge most often refers to wire size, which affects resistance, current handling, and signal integrity. Lower gauge numbers indicate thicker wire with lower resistance. Choosing the correct gauge ensures reliable performance in loudspeaker and power applications.

Geometrical Acoustics — A branch of acoustics that models sound propagation using rays rather than waves. It assumes wavelengths are small relative to room dimensions, allowing reflections, refractions, and scattering to be treated like light. This approach is useful for architectural acoustics and large-scale modeling. It simplifies complex environments while providing practical design insight.

Gooseneck — A flexible, positionable support arm used to mount microphones or lights. Its segmented design allows precise placement without tools. Goosenecks are common in podiums, broadcast setups, and instrument miking. Their stability and adjustability make them ideal for spoken-word applications.

Grain — A subtle, textured quality in audio that may arise from noise, distortion, or digital artifacts. Grain can make sound feel rough or gritty, especially in high frequencies. Some listeners find it unpleasant, while others associate it with vintage character. Reducing grain improves clarity and smoothness.

Graphic Equalizer — An equalizer with fixed frequency bands controlled by sliders arranged in a visual "graph." Each slider boosts or cuts a specific band, allowing intuitive tonal shaping. Graphic EQs are common in live sound, installed systems, and room tuning. Their layout provides quick, broad control over frequency balance.

Graphic-Level Recorder — A device that visually records sound levels over time, often using a moving pen or digital display. It provides a long-term view of noise patterns, peaks, and trends. These recorders support environmental studies, compliance testing, and system monitoring. Their graphical output simplifies interpretation.

Grating Diffraction — A diffraction effect produced when sound encounters a periodic structure such as slats or perforations. The spacing of the grating determines how sound is scattered or redirected. Grating diffraction influences acoustic treatments, architectural features, and measurement setups. It can enhance diffusion or create unwanted artifacts.

Grating Reflection Phase — The phase shift introduced when sound reflects from a periodic grating structure. This phase behavior affects interference patterns and directional scattering. Understanding it helps design diffusers and acoustic surfaces. Proper control improves spatial uniformity and reduces coloration.

Green's Function — A mathematical tool used to describe how a system responds to an impulse at a specific location. In acoustics, it models sound propagation, reflections, and boundary interactions. Green's functions support advanced simulation and inverse problems. They form the foundation of many computational acoustics methods.

Ground — A reference point in an electrical system used for safety and noise control. Proper grounding prevents shock hazards and stabilizes circuit operation. In audio, grounding helps reduce hum and interference. Good grounding practices are essential for clean signal paths.

Ground Loop — An unwanted electrical loop created when multiple devices share different ground paths. Ground loops introduce hum, buzz, or interference into audio systems. They often occur in complex setups with multiple power sources. Solutions include isolation transformers, balanced connections, and proper grounding.

Group Velocity — The speed at which energy or information travels through a medium in a wave packet. Group velocity differs from phase velocity and determines how signals propagate in dispersive environments. It is important in room acoustics, waveguides, and material analysis. Understanding it helps predict arrival times and energy flow.

Haas Effect — A psychoacoustic phenomenon where early reflections arriving within about 30 ms of the direct sound are fused into a single auditory event. These reflections enhance loudness and spatial impression without causing echo. The Haas effect guides loudspeaker placement and room design. It is central to perceived clarity and localization.

Hair Cell — A sensory cell in the cochlea that converts mechanical vibrations into neural signals. Hair cells are frequency-specific and arranged tonotopically along the basilar membrane. Damage to them leads to permanent hearing loss. They are essential for pitch perception and auditory detail.

Handheld — Referring to microphones or devices designed to be held in the hand during use. Handheld mics are common in live performance, interviews, and public speaking. Their design emphasizes durability, handling comfort, and feedback resistance. Proper technique minimizes noise and improves clarity.

Handling Noise — Unwanted sound generated when a microphone is touched, moved, or vibrated. It results from mechanical coupling between the mic body and capsule. Handling noise can obscure speech or musical detail. Shock mounts and proper technique reduce its impact.

Hard Room — A room with many reflective surfaces and little absorption. Hard rooms produce strong reflections, long reverberation, and potential intelligibility issues. They emphasize brightness and can cause flutter echo. Acoustic treatment helps balance their sound.

Harmonic — A frequency component that is an integer multiple of a fundamental frequency. Harmonics shape timbre and contribute to the richness of musical tones. Their relative levels define the character of instruments and voices. Harmonics are central to both acoustics and signal analysis.

Harmonic Distortion — Distortion that introduces harmonic frequencies not present in the original signal. It results from nonlinear behavior in equipment or transducers. Low levels may be inaudible or even musically pleasing; high levels degrade clarity. Harmonic distortion is a key metric in audio quality evaluation.

Harmonics — The set of integer-multiple frequency components that accompany a fundamental tone. Harmonics determine timbre, warmth, and brightness. Their distribution varies across instruments and voices. Understanding harmonics is essential for synthesis, mixing, and acoustic analysis.

Hash — A type of broadband noise characterized by a rough, gritty texture. Hash often results from electrical interference, digital artifacts, or faulty equipment. It can mask detail and reduce audio quality. Identifying the source is key to eliminating it.

Head — In audio, "head" may refer to the magnetic heads in tape machines or the top portion of a drum. Tape heads record and play back magnetic signals, requiring precise alignment. Drum heads determine tone, resonance, and attack. The term's meaning depends on context.

Headphones — Personal listening devices that place transducers on or around the ears. Headphones provide isolation, detail, and portability for monitoring and enjoyment. Their design influences comfort, frequency response, and spatial imaging. They are essential tools in production and everyday listening.

Head-Related Impulse Response — A measurement of how sound arrives at the ears from a specific direction, capturing reflections, diffraction, and body effects. HRIRs form the basis of binaural audio and spatial simulation. They encode precise localization cues. Accurate HRIRs enable immersive headphone experiences.

Head-Related Transfer Function — A frequency-domain representation of how the head, torso, and ears shape incoming sound. HRTFs describe direction-dependent filtering essential for localization. They are used in virtual acoustics, gaming, and spatial audio rendering. Individual HRTFs vary significantly between listeners.

Headroom — The margin between normal operating level and the point of distortion or clipping. Adequate headroom ensures clean reproduction of peaks and transients. Systems with limited headroom distort more easily. Proper gain staging preserves headroom and audio quality.

Head-Worn — Referring to microphones or devices worn on the head, such as headset mics or in-ear monitors. Head-worn gear provides consistent placement and hands-free operation. It is widely used in theater, broadcast, and live performance. Proper fit ensures comfort and stability.

Hearing Impaired — Describing individuals with reduced hearing ability due to medical, genetic, or environmental factors. Hearing impairment ranges from mild to profound and affects communication and perception. Assistive technologies and acoustic design help improve accessibility. The term is used in clinical and regulatory contexts.

Hearing Loss — A reduction in hearing sensitivity caused by aging, noise exposure, disease, or injury. Hearing loss may affect specific frequencies or overall perception. It impacts communication, safety, and quality of life. Prevention and early diagnosis are essential for long-term hearing health.

Hearing Protector — A device designed to reduce sound exposure and prevent hearing damage. Common types include earplugs and earmuffs. Proper fit and usage determine effectiveness. Hearing protectors are essential in industrial, recreational, and military environments.

Hearing Sensitivity — The ear's ability to detect faint sounds across different frequencies. Sensitivity varies with age, health, and environmental conditions. It is highest in the mid-frequency range important for speech. Measuring sensitivity helps diagnose hearing conditions.

Hearing Threshold — The lowest sound level a listener can detect at a given frequency. Thresholds form the basis of audiograms and hearing assessments. They vary across individuals and change with age or exposure. Threshold shifts indicate potential hearing damage.

Helmholtz Number — A dimensionless parameter comparing the size of an acoustic cavity to the wavelength of sound. It helps predict resonance behavior and flow interactions. Helmholtz numbers are used in modeling ducts, resonators, and aerodynamic noise. They guide design decisions in acoustics and fluid dynamics.

Helmholtz Resonator — A cavity with a narrow neck that resonates at a specific frequency when excited by sound. Helmholtz resonators absorb or amplify targeted frequencies depending on design. They are used in room acoustics, mufflers, and instrument construction. Their behavior is governed by cavity volume and neck geometry.

Hemi-Anechoic Room — A room designed to be reflection-free on walls and ceiling but with a reflective floor. This creates a half-space acoustic environment suitable for testing loudspeakers, machinery, and vehicles. Hemi-anechoic rooms simulate outdoor ground-plane conditions. They balance realism with controlled measurement.

Henry — The SI unit of inductance, symbolized as H. Inductors measured in henries store energy in magnetic fields and influence frequency response. In audio circuits, inductance affects filters, crossovers, and transformers. Understanding henries helps design stable, predictable systems.

Hertz — The unit of frequency equal to one cycle per second. Hertz describes how often a sound wave repeats and directly determines perceived pitch. Low hertz values correspond to bass, while high values correspond to treble. It is the foundational measurement for all acoustic and audio frequency analysis.

High — A general descriptor for the upper portion of the audio spectrum, typically above 2 kHz. High frequencies contribute brightness, clarity, and detail to sound. Excessive highs can cause harshness, while insufficient highs make audio dull. Managing high-frequency balance is essential in mixing and system tuning.

High Cut Filter — A filter that reduces high-frequency content while allowing lower frequencies to pass. It is used to soften harshness, remove hiss, or shape tonal character. High-cut filters are common in mixing, mastering, and broadcast processing. Their slope determines how aggressively highs are reduced.

High Pass Filter — A filter that allows high frequencies to pass while attenuating lower frequencies. High-pass filters remove rumble, plosives, and low-frequency noise. They are essential in microphone processing, system tuning, and audio cleanup. Proper use improves clarity and headroom.

High-Band UHF — A portion of the ultra-high-frequency spectrum used for wireless microphones and communication devices. High-band UHF offers good range, low interference, and reliable performance. Its availability varies by region due to spectrum regulations. Proper frequency coordination ensures stable operation.

High-Band VHF — A segment of the very-high-frequency spectrum used for wireless audio and broadcast applications. High-band VHF provides longer wavelength propagation and can penetrate obstacles better than higher bands. It is more susceptible to noise but remains useful in certain environments. Frequency planning is essential for interference-free use.

Hiss — A broadband, high-frequency noise often caused by electronic circuitry, poor gain staging, or tape systems. Hiss can mask detail and reduce perceived fidelity. It is especially noticeable in quiet passages. Proper equipment maintenance and noise reduction techniques help minimize hiss.

Horn Loading — A method of improving loudspeaker efficiency by coupling the driver to a horn-shaped structure. Horn loading increases acoustic output and controls directivity. It is widely used in PA systems and high-efficiency designs. Proper horn geometry ensures smooth response and reduced distortion.

Hot — A term describing a signal level that is unusually strong or close to clipping. A hot signal can improve signal-to-noise ratio but risks distortion if not controlled. Engineers monitor hot levels carefully during recording and mixing. Balanced gain staging prevents overload.

House — In live sound, "house" refers to the main loudspeaker system that delivers audio to the audience. The house system must provide even coverage, clarity, and appropriate loudness. It is distinct from monitor systems used by performers. Proper tuning ensures a consistent listening experience.

Hum — A low-frequency noise, often at 50 or 60 Hz, caused by electrical interference or grounding issues. Hum can severely degrade audio quality and distract listeners. It commonly arises from ground loops or poorly shielded cables. Eliminating hum requires proper grounding and isolation.

Hydrophone — An underwater microphone designed to detect sound in liquids. Hydrophones are used in marine research, underwater communication, and environmental monitoring. Their design accounts for pressure, impedance, and waterborne propagation. They reveal acoustic phenomena not present in air.

Hydrophone Directivity — The directional sensitivity pattern of a hydrophone. Directivity determines how well it captures sound from specific angles underwater. Patterns vary from omnidirectional to highly focused designs. Proper directivity selection improves measurement accuracy and target detection.

Hypercardioid Microphone — A highly directional microphone pattern with a narrow front pickup and a small rear lobe. Hypercardioid mics offer excellent isolation and feedback resistance. They are ideal for stage use, film production, and noisy environments. Placement must account for the rear pickup zone.

Hypersonic — Referring to extremely high frequencies beyond the upper limit of human hearing, often above 20 kHz. Hypersonic content may influence perception indirectly through nonlinearities or tactile effects. It is studied in psychoacoustics and high-resolution audio. Most playback systems do not reproduce hypersonic frequencies.

Hypersonic Flow — A fluid flow regime where speeds exceed Mach 5. In acoustics, hypersonic flow affects shock waves, noise generation, and aerodynamic heating. It is relevant in aerospace engineering and high-speed vehicle design. Modeling hypersonic flow requires advanced computational methods.

Hypersonic Glider — A vehicle designed to travel at hypersonic speeds while gliding through the atmosphere. Its acoustic signature includes intense shock waves and aerodynamic noise. Hypersonic gliders pose unique challenges for noise prediction and environmental impact. Their design integrates aerodynamics, materials science, and acoustics.

Hypersonics — The study of phenomena associated with extremely high-speed motion, typically above Mach 5. Hypersonics includes shock formation, aerodynamic noise, and thermal effects. It intersects with aeroacoustics and fluid dynamics. Research in hypersonics supports advanced aerospace technologies.

Hz — The abbreviation for hertz, the standard unit of frequency. It indicates how many cycles per second a sound wave completes. Hz values define pitch, tonal balance, and system bandwidth. It is used universally in audio and acoustics.

I — A symbol that may represent electric current, intensity, or other variables depending on context. In acoustics, "I" often denotes sound intensity, the power per unit area. In electronics, it represents current measured in amperes. Interpretation depends on the surrounding notation.

IEC 61672 — An international standard defining performance requirements for sound level meters. It specifies accuracy classes, frequency weighting, time weighting, and calibration procedures. Compliance ensures reliable and comparable noise measurements. The standard is widely used in environmental and occupational acoustics.

IEEE — The Institute of Electrical and Electronics Engineers, a global organization that develops technical standards and supports engineering research. IEEE standards influence audio, communication, and signal processing technologies. The organization publishes journals and hosts conferences across many disciplines. Its work shapes modern electronic and acoustic systems.

Imaging — The ability of an audio system to create a precise spatial representation of sound sources. Good imaging allows listeners to perceive direction, distance, and separation between instruments. It depends on speaker placement, room acoustics, and recording techniques. Strong imaging enhances realism and listener engagement.

Impact Insulation Class — A rating that measures a floor-ceiling assembly's ability to reduce impact noise such as footsteps. Higher IIC values indicate better isolation. It is essential in multi-story buildings and residential construction. Field performance may differ from laboratory ratings due to flanking paths.

Impact Sound Pressure Level — The sound pressure level generated by an impact on a floor or structure. It is measured using standardized tapping machines. Lower levels indicate better impact noise control. The metric helps evaluate flooring systems and building assemblies.

Impact Sound Transmission — The transfer of impact-generated noise through building structures. It includes footsteps, dropped objects, and mechanical vibrations. Effective control requires resilient materials, decoupling, and proper construction. Impact transmission is a major factor in building acoustics.

Impedance — The opposition a system presents to the flow of alternating current or sound energy. Impedance includes both resistive and reactive components. Matching impedance between devices ensures efficient energy transfer. It is fundamental in loudspeaker, microphone, and circuit design.

Impedance Curve — A graph showing how a loudspeaker's impedance varies with frequency. Peaks and dips reveal resonances, enclosure effects, and driver behavior. The curve guides crossover design and amplifier matching. Understanding it prevents overload and improves performance.

Impedance Matching — The process of aligning the impedance of two devices or systems to maximize energy transfer and minimize reflections. Proper matching improves efficiency, clarity, and stability. It is essential in microphones, loudspeakers, and transmission lines. Mismatched impedance can cause distortion or loss of level.

Impedance Tube — A measurement device used to determine the acoustic impedance and absorption of materials. It uses standing waves in a cylindrical tube to analyze reflection and transmission. Impedance tubes provide precise, frequency-dependent data. They are widely used in material research and product development.

Impulse — A very short, high-energy sound event used to excite acoustic systems. Impulses reveal resonances, reflections, and system response. They are used in measurement, modeling, and testing. Impulse signals form the basis of impulse response analysis.

Impulse Invariance — A digital filter design method that preserves the time-domain impulse response of an analog filter. It ensures similar behavior between analog and digital versions at low frequencies. However, it may introduce aliasing at higher frequencies. The method is used in certain DSP applications.

Impulse Response — A recording of how a system reacts to an impulse, capturing reflections, resonances, and decay. Impulse responses characterize rooms, loudspeakers, and devices. They are essential for convolution reverb, acoustic modeling, and system tuning. IRs provide a complete snapshot of acoustic behavior.

Impulsive Noise — Noise consisting of sudden, sharp bursts such as bangs, clicks, or impacts. Impulsive noise can be more damaging and disturbing than continuous noise. It requires specialized measurement techniques and protective strategies. It is common in industrial, construction, and environmental settings.

In Phase — A condition where two signals have matching timing and waveform alignment. In-phase signals reinforce each other, increasing amplitude. Out-of-phase signals can cancel or weaken. Phase alignment is crucial in multi-mic setups and loudspeaker arrays.

Inductance — The property of an electrical component that resists changes in current by storing energy in a magnetic field. Inductance affects filters, crossovers, and transformer behavior. It is measured in henries and influences frequency response. Proper inductance control ensures stable, predictable audio circuits.

Infrared — Electromagnetic radiation with wavelengths longer than visible light, commonly used for wireless remote controls and communication links. In audio, infrared systems transmit signals for assistive listening and silent-conference applications. They offer immunity to RF interference but require line-of-sight. Infrared technology provides secure, localized signal distribution.

Infrasonic Frequency — Frequencies below the lower limit of human hearing, typically below 20 Hz. Infrasonic energy can be produced by natural events, machinery, or large-scale structures. Although inaudible, it may be felt physically or influence perception indirectly. Infrasonic analysis is important in environmental and structural acoustics.

Infrasonic Sound — Sound waves with frequencies below 20 Hz that humans generally cannot hear. These waves travel long distances and penetrate structures easily. Infrasonic sound can cause vibration, discomfort, or physiological effects at high levels. It is studied in seismology, meteorology, and industrial noise control.

Initial Time-Delay Gap — The time interval between the arrival of direct sound and the first significant reflection. A short gap enhances clarity, while a long gap can create a sense of spaciousness or echo. The gap is a key parameter in room acoustics and speech intelligibility. Designers adjust surfaces and geometry to optimize this interval.

INM — A shorthand for *Inter-Noise Masking*, referring to the phenomenon in which one sound (the masker) reduces the audibility or clarity of another sound (the target) due to overlapping spectral content, temporal proximity, or spatial interaction. In acoustics and psychoacoustics, INM helps explain why certain frequencies become obscured in noisy rooms, why speech intelligibility drops in reverberant or broadband-noise environments, and why system designers must manage both spectral balance and temporal smear. It is central to understanding masking curves, critical bands, and perceptual thresholds in real-world listening spaces.

Input — The point at which an audio signal enters a device or system. Inputs may accept microphone, line, instrument, or digital signals depending on design. Proper input selection ensures correct gain structure and noise performance. Inputs form the starting point of the signal chain.

Input Attenuator — A control or device that reduces the level of a signal before it enters a circuit. Attenuators prevent overload, distortion, and clipping in sensitive equipment. They are common in mixers, preamps, and measurement devices. Proper attenuation preserves headroom and clarity.

Insert — A connection point on a mixer that allows external processors to be placed directly into a channel's signal path. Inserts support compressors, EQs, and other inline effects. They provide precise control over individual channels. Insert jacks typically use tip-send, ring-return wiring.

Insertion Gain Loss — The change in signal level caused by inserting a device into a signal path. Ideally, insertion should not alter level, but real devices may introduce gain or loss. Measuring insertion effects ensures proper gain staging. Excessive loss may indicate impedance mismatch or faulty equipment.

Insertion Loss — The reduction in signal level caused by adding a component, filter, or device into a system. In acoustics, it also refers to the noise reduction achieved by installing a barrier or silencer. Lower insertion loss indicates better performance. It is a key metric in both audio and noise control.

Instantaneous Sound Pressure — The moment-to-moment variation in air pressure caused by a sound wave. These fluctuations form the raw waveform before averaging or weighting. Instantaneous pressure determines waveform shape and peak levels. It is fundamental to understanding sound behavior.

Instrument Level — A signal level produced by electric guitars, basses, and some keyboards, typically lower than line level and higher than microphone level. Instrument-level signals require proper impedance matching and gain to avoid noise or tone loss. Direct boxes and preamps convert them for mixing. Correct handling preserves dynamics and character.

Integrating-Averaging Sound Level Meter — A sound level meter that measures fluctuating noise by integrating sound energy over time. It provides accurate long-term averages such as Leq. These meters are essential for environmental and occupational noise assessments. They capture realistic exposure levels beyond instantaneous readings.

Intensity — The sound power passing through a unit area, typically measured in watts per square meter. Intensity describes the directional flow of acoustic energy. It is used in noise source identification and sound field analysis. Intensity differs from sound pressure but is closely related.

Intensity Probe — A measurement device that uses two closely spaced microphones to determine sound intensity and direction. Intensity probes reveal energy flow, reflections, and source locations. They are valuable in noise diagnostics and acoustic research. Proper calibration ensures accurate results.

Interference — The interaction of two or more sound waves that combine to produce reinforcement or cancellation. Constructive interference increases amplitude, while destructive interference reduces it. Interference patterns shape room acoustics, microphone behavior, and loudspeaker arrays. Managing interference improves clarity and consistency.

Interference Tube — A long tube used in shotgun microphones to create directional patterns through phase cancellation. The tube's slots allow off-axis sound to enter at different times, reducing its level. This design yields strong forward focus and rejection. Interference tubes are essential in film and broadcast audio.

Intermittent Noise — Noise that starts and stops irregularly, such as traffic bursts or machinery cycles. Intermittent noise is often more disturbing than steady noise due to unpredictability. It requires specialized metrics for accurate assessment. Environmental regulations often address intermittent sources separately.

Intermodulation — The creation of new frequencies when two or more signals interact in a nonlinear system. These frequencies are sums and differences of the originals. Intermodulation can cause distortion, interference, and reduced clarity. It is a key concern in wireless systems and audio electronics.

Intermodulation Distortion — Distortion caused by nonlinear mixing of multiple frequencies, producing unwanted sum and difference tones. IMD is often more objectionable than harmonic distortion because it is not harmonically related to the source. Low IMD indicates high-quality equipment. It is critical in amplifiers, wireless systems, and transducers.

Inverse-Square Law — A principle stating that sound pressure level decreases by 6 dB each time distance from a point source doubles in a free field. This law applies when reflections are minimal and the source radiates uniformly. It helps predict coverage and level distribution. Deviations occur in reverberant or directional environments.

IR — An *impulse response*, the complete acoustic fingerprint of a space, system, or device captured by measuring how it reacts to a brief, broadband excitation. It encodes all reflections, resonances, absorption characteristics, and temporal behaviors that shape what a listener ultimately hears. In room acoustics, IRs are used to derive reverberation time, clarity metrics, early-reflection structure, and frequency-dependent decay. In DSP, they serve as the basis for convolution, system identification, and filter design, allowing engineers to recreate or analyze acoustic environments with precision.

ISO 3382 Parameters — A standardized set of room acoustic metrics including reverberation time, clarity, definition, and early decay time. These parameters describe how sound behaves in performance spaces and auditoriums. They guide design, evaluation, and comparison of rooms. ISO 3382 is widely used in architectural acoustics.

Isobar — A line on a graph connecting points of equal pressure. In acoustics, isobars illustrate sound pressure distribution in a field. They help visualize patterns, hotspots, and nulls. Isobar mapping supports system tuning and room analysis.

Isolate — To separate a sound source or component to prevent unwanted transmission or interference. Isolation may involve physical decoupling, damping, or enclosure design. Effective isolation improves clarity and reduces noise problems. It is essential in studios, buildings, and mechanical systems.

Isolated Burst of Sound Energy — A short, discrete sound event separated from other noise by silence. These bursts are used in testing, measurement, and psychoacoustic research. They reveal transient behavior and system response. Their clarity makes them useful for diagnostic work.

Isolation — The reduction of sound transmission between spaces or components. Isolation relies on mass, decoupling, and airtight construction. Good isolation improves privacy, comfort, and recording quality. It is a core principle of building and studio acoustics.

Isolation Transformer — A transformer that electrically separates input and output circuits to prevent ground loops and interference. It provides clean, noise-free signal transfer in audio systems. Isolation transformers are common in live sound, broadcast, and measurement setups. They enhance safety and signal integrity.

ITD — *Interaural Time Difference*, the tiny difference in arrival time between a sound reaching the left and right ears, serving as one of the primary cues the auditory system uses to localize sources on the horizontal plane. Low-frequency sounds produce the most salient ITD cues because their long wavelengths allow the brain to track phase-coherent timing differences with high precision. In room acoustics and spatial audio design, ITD informs panning, binaural rendering, and the perception of lateral envelopment, making it essential for accurate imaging and immersive playback.

Jack — A connector or receptacle used to interface audio cables and equipment. Common types include TRS, TS, and XLR jacks. Jacks provide secure, standardized connections for signals. Their quality affects reliability and noise performance.

JAES — The *Journal of the Audio Engineering Society*, a peer-reviewed publication that disseminates research on acoustics, psychoacoustics, signal processing, transducer design, spatial audio, measurement techniques, and emerging audio technologies. It serves as a primary venue for foundational and cutting-edge work in room acoustics, loudspeaker behavior, binaural perception, DSP algorithms, and standards development. Because JAES bridges academic rigor with practical engineering relevance, it remains one of the most authoritative sources for advancing both theoretical understanding and applied practice in professional audio.

JASA — The *Journal of the Acoustical Society of America*, a leading peer-reviewed publication covering fundamental and applied research across acoustics, including architectural acoustics, psychoacoustics, signal processing, transducer design, underwater acoustics, noise control, and auditory perception. It serves as a primary venue for advancing theoretical models, measurement techniques, and experimental findings that shape modern acoustic science and engineering practice. Because JASA bridges physics, engineering, and perceptual research, it remains one of the most influential sources for scholars, system designers, and practitioners seeking authoritative, data-driven insight into sound and hearing.

Johnson–Nyquist Noise — The thermal noise generated by the random motion of electrons in a resistor. It is present in all electronic circuits and sets a fundamental noise floor. Johnson–Nyquist noise increases with temperature and bandwidth. Understanding it helps design low-noise audio equipment.

Joist — A structural beam used to support floors or ceilings. Joists can transmit sound through buildings, contributing to flanking noise. Proper isolation and construction reduce unwanted transmission. Joist behavior is important in architectural acoustics.

kHz — A unit equal to one thousand cycles per second, used to express audio and acoustic frequencies in the mid- to high-frequency range where human hearing is most sensitive. Many critical perceptual phenomena—such as speech intelligibility, sibilance, localization cues, and harmonic detail—occur in the 1–10 kHz region, making kHz-scale measurements central to loudspeaker design, microphone response, room-acoustic analysis, and equalization work.

Kirchhoff Integral — A mathematical formulation used to describe wave propagation and diffraction based on boundary conditions. In acoustics, it helps model how sound interacts with surfaces and apertures. The Kirchhoff integral underpins many computational and analytical methods. It provides insight into complex sound fields.

Kirchhoff–Helmholtz Equation — A fundamental integral equation describing how sound fields behave in a region based on boundary values. It relates pressure and particle velocity on a surface to the resulting sound field in space. The equation underpins many numerical methods in acoustics, including boundary element modeling. It provides a rigorous framework for predicting reflections, scattering, and radiation.

Knee — The point on a compressor's transfer curve where gain reduction begins. A "hard knee" introduces compression abruptly, while a "soft knee" applies it gradually for smoother results. Knee behavior strongly influences transparency and character. Engineers adjust it to shape dynamics naturally.

Korner Killer — A type of acoustic treatment designed to absorb low-frequency energy in room corners. It reduces standing waves, boominess, and modal buildup. Korner Killers improve clarity and bass definition in studios and listening rooms. Their placement in trihedral corners maximizes effectiveness.

Kurtosis — A statistical measure describing the "peakedness" or tail heaviness of a distribution. In acoustics, kurtosis helps characterize noise with strong impulsive components. High kurtosis indicates sharp, transient-rich signals that may be more damaging to hearing. It is increasingly used in advanced noise exposure assessment.

K-Weighting — A frequency weighting curve used in modern loudness measurement standards such as ITU-R BS.1770. It reflects human sensitivity to broadband program material rather than pure tones. K-weighting supports consistent loudness normalization across media. It is central to broadcast and streaming workflows.

Lagrangian Acoustics — An approach to acoustics that tracks individual fluid particles rather than analyzing fields at fixed points. It is useful for modeling nonlinear effects, turbulence, and particle-based phenomena. Lagrangian methods complement traditional Eulerian approaches. They support advanced simulation of complex acoustic environments.

Lapel Microphone — A small microphone clipped to clothing near the speaker's chest. Lapel mics provide hands-free operation and consistent vocal pickup. They are widely used in broadcast, theater, and presentations. Proper placement minimizes clothing noise and improves clarity.

Laplacian Speed of Sound — A theoretical expression for sound speed derived from the Laplace equation under idealized conditions. It applies to inviscid, adiabatic media where thermodynamic relationships are simplified. The concept helps illustrate fundamental wave behavior. Real-world sound speed varies with temperature, humidity, and medium properties.

LAR — *Loudness Adjustment Range*, the span over which a playback or monitoring system can vary perceived loudness while maintaining stable tonal balance, low distortion, and consistent spatial imaging. In acoustics and system calibration, LAR reflects how well a system preserves spectral integrity and dynamic behavior as levels change, ensuring that mixes, measurements, and listening evaluations remain reliable at both low and high playback volumes. A wide, well-controlled LAR is essential for critical listening environments, where engineers must judge balance, detail, and spatial cues across a range of monitoring levels.

Lav — Short for "lavalier microphone," another name for a lapel microphone. Lav mics are compact, discreet, and ideal for on-camera use. They provide consistent vocal pickup with minimal visual distraction. Proper cable management reduces rustle and handling noise.

Law of the First Wavefront — A psychoacoustic principle stating that the first arriving sound determines perceived direction, even if later reflections are louder. This effect supports localization in reverberant spaces. It guides loudspeaker placement and room design. Early reflections must be controlled to preserve directional clarity.

LCR — A three-channel loudspeaker configuration consisting of Left, Center, and Right front channels, used to anchor imaging, stabilize dialogue or primary content, and maintain consistent localization across a wide listening area. In acoustics and sound-system design, LCR layouts reduce phantom-center instability, improve coverage uniformity, and allow mixers to distribute spectral and spatial information more precisely than in simple stereo. The center channel provides a dedicated, position-independent reference point, while the left and right channels supply width, envelopment, and directional cues.

Leakage — Unwanted sound entering a microphone or channel from nearby sources. Leakage reduces isolation and complicates mixing. It is influenced by microphone pattern, placement, and room acoustics. Managing leakage improves clarity and control.

LEDE — *Live End–Dead End*, an acoustic design philosophy for control rooms in which the front portion of the room (the "dead end") is heavily absorptive to minimize early reflections, while the rear portion (the "live end") is reflective or diffusive to preserve spaciousness and natural decay. This arrangement improves imaging, reduces comb-filtering, and stabilizes the stereo soundstage by ensuring that the listener receives a clean direct signal from the monitors followed by controlled, non-destructive ambience. LEDE rooms were foundational in modern studio design, influencing later concepts such as RFZ and non-environment approaches.

Level — The magnitude of an audio signal, typically expressed in decibels. Level determines loudness, headroom, and gain structure. Maintaining proper levels ensures clean, distortion-free audio. Level management is central to all audio workflows.

Level Above Threshold — The amount by which a signal exceeds a defined threshold in dynamics processing. It determines how aggressively compressors or limiters act. Higher levels above threshold trigger stronger gain reduction. This metric shapes the character of dynamic control.

Level Reduction — The decrease in signal amplitude caused by processing, attenuation, or system behavior. Level reduction may be intentional, as in compression, or unintentional due to loss or mismatch. Monitoring reduction ensures proper gain staging. Excessive reduction can degrade clarity.

Level Sensor — A device or circuit that detects signal level for control or monitoring purposes. Level sensors trigger automation, protection circuits, or dynamic processing. They support stable operation in audio systems. Accurate sensing ensures predictable behavior.

LFD — *Low-Frequency Directivity*, the degree to which a loudspeaker or array controls the spread of bass energy, typically below a few hundred hertz where wavelengths are long and most sources radiate nearly omnidirectionally. In acoustics, improving LFD helps reduce room-mode excitation, tighten bass imaging, and minimize low-frequency spill into adjacent spaces. Techniques such as cardioid subwoofer arrays, gradient configurations, boundary placement, and enclosure geometry are used to shape LFD and achieve more predictable, controlled low-frequency behavior in both live-sound and studio environments.

Lighthill's Acoustic Analogy — A theoretical framework describing how turbulent fluid motion generates sound. It forms the basis of aeroacoustics and explains noise from jets, fans, and high-speed flows. Lighthill's analogy models sound as a byproduct of fluid stress and vorticity. It is foundational in computational aeroacoustics.

Limiter — A dynamics processor that prevents a signal from exceeding a set maximum level. Limiters protect equipment, prevent clipping, and control peaks. They are essential in live sound, broadcast, and mastering. Proper settings maintain loudness while avoiding distortion.

Limiting — The process of restricting a signal's maximum amplitude using a limiter. Limiting ensures peaks remain within safe or desired bounds. It can increase perceived loudness but may reduce dynamics if overused. Transparent limiting requires careful threshold and release settings.

Line Level — A standardized signal level used for interconnecting audio equipment. Line level is stronger than microphone level but weaker than speaker level. It ensures compatibility between mixers, processors, and interfaces. Proper matching avoids noise and distortion.

Line Loss — The reduction in signal strength as it travels through a cable or transmission line. Loss increases with distance, frequency, and cable resistance. Excessive line loss degrades clarity and level. High-quality cables and proper impedance matching minimize it.

Line Spectrum — A spectrum consisting of discrete frequency components rather than a continuous range. Line spectra occur in tonal sounds, harmonics, and periodic signals. They reveal structure and pitch relationships. Spectrum analysis helps identify line components.

Linear — Describing a system whose output is directly proportional to its input. Linear behavior preserves waveform shape and avoids distortion. Many audio processes aim for linearity to maintain fidelity. Nonlinear systems introduce harmonics or compression.

Lisp — A speech artifact where sibilant sounds become smeared, softened, or distorted. Lisping may result from microphone placement, processing errors, or vocal characteristics. Excessive de-essing can also cause lisping. Proper technique preserves natural articulation.

Live — Describing a room or sound with strong reflections and long reverberation. Live spaces feel lively, bright, and spacious. They enhance musical envelopment but may reduce speech clarity. Balancing liveliness with control is key in acoustic design.

Live End Dead End — An acoustic design strategy where one end of a room is absorptive (dead) and the opposite end is reflective or diffusive (live). This arrangement improves imaging and reduces early reflections while maintaining natural ambience. It is widely used in control rooms. Proper implementation requires careful material selection.

Live Room — A recording space with reflective surfaces and moderate to long reverberation. Live rooms add natural ambience and energy to recordings. They are ideal for drums, ensembles, and acoustic instruments. Their character contrasts with dry, isolated booths.

Load — The electrical or acoustic impedance presented to a device or system. In audio, load affects amplifier performance, speaker behavior, and signal transfer. Proper matching ensures efficiency and prevents damage. Incorrect loading can cause distortion or instability.

Lobe — A region of increased sensitivity or radiation in a microphone or loudspeaker's polar pattern. Lobes reveal directional behavior and off-axis response. Secondary lobes may cause coloration or feedback issues. Understanding lobes aids in placement and system design.

Logarithm — A mathematical function that expresses ratios on a compressed scale. Logarithms underpin the decibel system, frequency perception, and many acoustic formulas. They allow large ranges to be represented compactly. Logarithmic thinking is fundamental to audio engineering.

Lombard Effect — The involuntary tendency for speakers to increase vocal level in noisy environments. The effect helps maintain intelligibility but can strain the voice. It influences communication, public address design, and noise studies. Understanding it supports better acoustic environments.

Longitudinal Wave — A wave in which particles oscillate parallel to the direction of propagation. Sound in air and liquids travels primarily as longitudinal waves. These waves alternate between compression and rarefaction. Their behavior defines fundamental acoustic properties.

Loudness — The subjective perception of sound intensity influenced by frequency, duration, and context. Loudness does not always correlate directly with sound pressure level. Psychoacoustic models and weighting curves help quantify it. Loudness perception guides mixing, mastering, and noise regulation.

Loudness Level — A psychoacoustic measure expressing perceived loudness in phons. It compares a sound's loudness to that of a pure tone at 1 kHz. Loudness level accounts for the ear's frequency-dependent sensitivity. It is essential in evaluating how humans actually experience sound intensity.

Loudspeaker — A transducer that converts electrical signals into audible sound through diaphragm motion. Loudspeakers vary in size, design, and frequency range, from compact tweeters to large subwoofers. Their performance depends on driver quality, enclosure design, and crossover integration. They are central to all playback and reinforcement systems.

Low Band UHF — A portion of the ultra-high-frequency spectrum used for wireless microphones and communication devices. Low-band UHF offers good range and moderate resistance to interference. Its availability depends on regional spectrum regulations. Proper coordination ensures stable wireless performance.

Low Band VHF — A segment of the very-high-frequency spectrum used for wireless audio and broadcast applications. Low-band VHF provides long-wavelength propagation and good obstacle penetration. It is more susceptible to noise but useful in certain environments. Frequency planning is essential for reliable operation.

Low Cut Filter — A filter that attenuates low-frequency content while allowing higher frequencies to pass. It removes rumble, plosives, and handling noise. Low-cut filters are common on microphones, mixers, and processors. Proper use improves clarity and reduces unwanted energy.

Low Pass Filter — A filter that allows low frequencies to pass while attenuating higher frequencies. It shapes tone, removes hiss, and supports crossover design. Low-pass filters are used in synthesis, mixing, and system tuning. Cutoff frequency determines the filter's tonal impact.

Lowlevel Noise Environment — A setting where background noise is extremely quiet, often below typical environmental levels.

Such environments allow precise acoustic measurements and critical listening. They are common in studios, test chambers, and research facilities. Maintaining low noise requires careful design and isolation.

Lumped Parameter Model — A simplified acoustic or mechanical model that treats distributed systems as discrete elements. It approximates complex behavior using masses, springs, and resistances. Lumped models are useful for low-frequency analysis and transducer design. They provide intuitive insight into system dynamics.

MAB — *Maximum Average Band*, a spectral-analysis metric that identifies the frequency band exhibiting the highest average energy over a defined time window. In acoustics, MAB is used to characterize dominant tonal regions, assess noise signatures, and compare how different environments or systems emphasize particular frequency ranges. It supports tasks such as room-response evaluation, loudness profiling, and spectral-balance tuning by highlighting where energy consistently concentrates across measurements.

Mach Number — The ratio of an object's speed to the speed of sound in the surrounding medium. Mach numbers describe flow regimes from subsonic to hypersonic. They influence shock waves, aerodynamic noise, and acoustic propagation. Mach effects are central to aeroacoustics.

Main Lobe — The primary region of strongest radiation or sensitivity in a polar pattern. The main lobe defines the direction in which a microphone or loudspeaker performs best. Its width and shape affect coverage and clarity. Understanding the main lobe guides placement and system design.

Main Speaker — The primary loudspeaker or loudspeaker array used to deliver sound to an audience. Main speakers must provide even coverage, clarity, and appropriate loudness. They form the core of a reinforcement system. Proper tuning ensures consistent performance across the listening area.

Mains — The primary AC power supply used to operate audio equipment. Mains voltage varies by region and must be matched to equipment specifications. Poor mains quality can introduce noise or instability. Proper grounding and conditioning ensure reliable operation.

Masked Threshold — The minimum level at which a sound becomes audible in the presence of masking noise. Masked thresholds vary with frequency and noise characteristics. They are essential in psychoacoustics, audio coding, and communication design. Understanding masking improves intelligibility and sound quality.

Masking — A psychoacoustic phenomenon where one sound makes another sound harder to hear. Masking depends on frequency, level, and temporal relationships. It influences mixing, hearing protection, and audio compression. Managing masking improves clarity and balance.

Masking Pattern — A graph showing how a masker affects audibility across frequencies. Masking patterns reveal how noise spreads its influence over nearby tones. They support psychoacoustic modeling and perceptual coding. Patterns vary with masker type and level.

Mass — A physical property describing the amount of matter in an object. In acoustics, mass influences vibration, resonance, and sound transmission. Heavier materials generally block sound more effectively. Mass is a key variable in building acoustics and transducer design.

Mass Law — A principle stating that sound transmission loss increases by approximately 6 dB for each doubling of surface mass. Mass law applies primarily to single, non-resonant panels. It guides material selection for sound isolation. Deviations occur at low frequencies and near resonances.

Mass Law Sound Transmission Loss — The sound reduction predicted by mass law for a given panel or barrier. It provides a baseline estimate of isolation performance. Real-world results may differ due to stiffness, damping, and flanking. Mass law remains a foundational concept in building acoustics.

Master — The primary control or output channel in an audio system. The master fader sets overall level for a mix or reinforcement system. Master channels may include processing such as EQ, limiting, or metering. They represent the final stage before amplification or recording.

Maximum A-Weighted Sound Pressure Level — The highest A-weighted sound level measured during an event. It reflects perceived loudness while accounting for human sensitivity. This metric is used in noise regulation, safety assessment, and product testing. It captures peak exposure more realistically than unweighted measures.

Maximum Response Axis — The direction in which a microphone or loudspeaker produces its strongest output. It defines the on-axis reference for frequency response measurements. Proper alignment with the maximum response axis ensures accurate performance. Off-axis behavior reveals dispersion characteristics.

Maximum Sound Level — The highest instantaneous sound pressure level recorded during a measurement period. It captures peak events such as impacts or bursts. Maximum levels are important in safety, environmental noise, and equipment testing. They differ from time-averaged metrics.

MDP Connector — A specialized connector type used in certain audio or communication systems. Its exact function depends on manufacturer and application. MDP connectors may carry power, data, or audio signals. Interpretation requires system context.

Mean Free Path — The average distance a sound particle or molecule travels before colliding with another. In acoustics, it relates to reverberation, diffusion, and energy decay. Longer mean free paths occur in large or sparsely reflective spaces. The concept supports statistical room acoustics.

Mean Square Pressure — The average of the squared instantaneous sound pressure values over time. It forms the basis for calculating sound pressure level. Mean square pressure smooths rapid fluctuations and represents acoustic energy. It is fundamental to measurement and analysis.

Measurement Uncertainty — The quantified doubt associated with a measurement result. It accounts for instrument accuracy, environmental conditions, and procedural variation. Understanding uncertainty ensures reliable interpretation of acoustic data. Standards require uncertainty reporting for compliance testing.

Mechanical Noise — Noise generated by moving parts in machinery, HVAC systems, or equipment. It includes vibration, rattling, and motor noise. Mechanical noise can interfere with recordings and reduce comfort. Isolation, damping, and maintenance help control it.

Megahertz — A unit equal to one million hertz. Megahertz values describe radio frequencies, digital clock rates, and high-frequency signals. They are used in wireless audio, DSP systems, and communication technologies. MHz is a standard unit in electronics and RF engineering.

Mel — A psychoacoustic unit representing perceived pitch. The mel scale maps frequency to human pitch perception more accurately than linear hertz. It is used in speech processing, auditory modeling, and filterbank design. The scale reflects how humans interpret pitch changes.

Meter — A device or display that measures audio level, loudness, or other parameters. Meters include VU, peak, RMS, and loudness meters. They guide gain staging, mixing, and compliance. Accurate metering ensures clean, consistent audio.

Metric Sabin — A unit of sound absorption equal to one square meter of perfectly absorbing surface. Metric sabins quantify how much sound energy a material removes from a room. They support reverberation time calculations and treatment design. The unit is central to architectural acoustics.

Microperforated Panel Absorber — An absorber that uses tiny perforations in a rigid panel to dissipate sound energy. These panels provide broadband absorption without fibrous materials. They are visually clean and suitable for architectural applications. Performance depends on perforation size, spacing, and cavity depth.

Microphone — A transducer that converts sound waves into electrical signals. Microphones vary in type, pattern, and sensitivity, each suited to specific applications. Their design influences clarity, noise performance, and tonal character. They are essential tools in recording, broadcast, and reinforcement.

Microphone Clip — A device used to secure a microphone to a stand or mount. Clips provide stability and proper positioning. Their design must match microphone size and shape. A good clip minimizes handling noise and ensures consistent placement.

Microphone Element — The core component of a microphone that performs the actual sound-to-electrical conversion. Elements may be dynamic, condenser, ribbon, or electret designs. Their construction determines sensitivity, frequency response, and durability. The element defines the microphone's fundamental character.

Microphone Level — A very low-level audio signal produced directly by a microphone, typically in the millivolt range. Because it is so weak, it requires significant amplification before further processing. Proper handling prevents noise and distortion. Microphone level is the starting point of most recording chains.

Microphone Preamp — A device that boosts microphone-level signals to usable line-level strength. Preamps influence noise performance, clarity, and tonal character. High-quality preamps provide clean gain with minimal coloration. They are essential in recording, broadcast, and live sound.

Microphone Sensitivity — A measure of how effectively a microphone converts sound pressure into electrical voltage. Higher sensitivity means stronger output for a given sound level. Sensitivity affects noise performance, gain requirements, and suitability for loud or quiet sources. It is a key specification in microphone selection.

Microphone Stand — A support structure used to position microphones securely and accurately. Stands come in boom, straight, and specialty designs for different applications. Stability and adjustability are essential for minimizing handling noise. Proper placement improves clarity and isolation.

Microphone Tree — A multi-microphone array used for orchestral and ensemble recording. Trees capture spacious, natural stereo imaging with balanced direct and ambient sound. Common designs include Decca Tree and modified variants. They are central to classical and film scoring techniques.

Middle Ear — The air-filled chamber between the eardrum and inner ear containing the ossicles. It amplifies and transmits sound vibrations to the cochlea. The middle ear equalizes pressure through the Eustachian tube. Its function is essential for efficient hearing.

MIDI — *Musical Instrument Digital Interface*, a communication protocol that transmits performance data—such as pitch, velocity, control changes, and timing—between electronic instruments, controllers, and software. Although it carries no audio, MIDI is central to acoustics-adjacent workflows because it enables precise control of synthesizers, samplers, virtual instruments, and automation systems used in sound design and measurement playback. Its event-based structure allows reproducible test signals, synchronized cueing, and detailed manipulation of parameters that influence timbre, dynamics, and spatial rendering in modern production and research environments.

Midrange — The middle portion of the audio spectrum, typically between 250 Hz and 2 kHz. Midrange frequencies carry most speech intelligibility and instrumental detail. Proper midrange balance is crucial for natural sound. Loudspeakers often use dedicated midrange drivers for accuracy.

Millisecond — One-thousandth of a second, a common unit for measuring delay, attack, and timing in audio. Millisecond values shape reverb, compression, and modulation effects. Small timing changes can significantly alter perception. Precision in milliseconds is essential for mixing and system alignment.

Mini Connector — A small audio connector, often referring to the 3.5 mm TRS plug used in consumer devices. Mini connectors carry stereo or microphone signals. They are compact but less robust than professional connectors. Proper handling prevents noise and wear.

Mix — The process of combining multiple audio tracks into a cohesive final presentation. Mixing balances levels, EQ, dynamics, and spatial placement. It shapes the emotional and sonic character of a production. A good mix enhances clarity, impact, and listener engagement.

Mix Down — The final stage of mixing where all elements are rendered into a stereo or multichannel master. Mixdowns consolidate processing, automation, and effects into a finished format. They prepare audio for mastering and distribution. Precision ensures translation across playback systems.

Mixer — A device or software tool that combines, routes, and processes audio signals. Mixers provide control over gain, EQ, dynamics, and effects. They are central to live sound, recording, and broadcast workflows. Mixers range from compact analog units to large digital consoles.

Mixing Console — A large, feature-rich mixer designed for professional audio production. Consoles offer extensive routing, automation, metering, and processing capabilities. They support complex sessions with many channels. Their layout promotes efficient workflow and precise control.

Modal Density — The number of resonant modes within a given frequency range in a room or structure. Higher modal density leads to smoother low-frequency response. Low density can cause uneven bass and strong resonances. It is a key concept in room acoustics.

Modal Overlap Factor — A measure of how much adjacent room modes overlap in frequency. High overlap produces smoother response, while low overlap creates distinct peaks and nulls. Overlap increases with room size and damping. It helps predict low-frequency behavior.

Modal Resonance — A resonance caused by standing waves in a room or structure. Modal resonances create peaks and dips in frequency response, especially at low frequencies. They depend on room dimensions and boundary conditions. Managing them improves clarity and tonal balance.

Mode — A natural resonance pattern in a room or object where standing waves form at specific frequencies. Modes create areas of reinforcement and cancellation. They are most problematic in small rooms. Understanding modes guides treatment and speaker placement.

Modulation — The process of varying one signal using another, affecting amplitude, frequency, or phase. Modulation creates effects such as tremolo, vibrato, and chorus. It is also fundamental to radio and wireless transmission. Modulation shapes both creative and technical audio processes.

Modulation Transfer Function — A measure of how well a system preserves modulation depth across frequencies. In acoustics, it relates to speech intelligibility and clarity. Higher MTF values indicate better transmission of detail. It is used in room analysis and communication system design.

Monaural — Single-channel audio reproduction without spatial separation. Monaural sound presents all content from one point, lacking stereo imaging. It is used in telephony, archival recordings, and certain broadcast contexts. Mono compatibility remains important in modern mixing.

Monitor — A loudspeaker or headphone used for critical listening during production. Monitors aim for accuracy, revealing flaws and details. Their placement and calibration affect translation across playback systems. Reliable monitoring is essential for professional results.

Monitor Bleed — Unwanted sound from stage monitors leaking into microphones. Bleed reduces isolation and complicates mixing. It is influenced by microphone pattern, placement, and monitor level. Managing bleed improves clarity and feedback stability.

Monitor Send — A signal routed from a mixer to a performer's monitor system. Monitor sends allow customized mixes for each performer. They may be pre- or post-fader depending on workflow. Proper monitor mixes enhance performance comfort and accuracy.

Monitor Speaker — A loudspeaker used for monitoring in studios or live sound. Monitor speakers prioritize accuracy, consistency, and low coloration. Their design supports critical decision-making. Placement and room treatment strongly influence performance.

Monophonic — Describing audio with a single channel or a synthesizer capable of playing one note at a time. Monophonic signals lack stereo width but can be powerful and focused. Mono synths are prized for expressive lead and bass lines. Mono playback compatibility remains important in mixing.

Monopole — A theoretical point source that radiates sound equally in all directions. Monopoles simplify acoustic modeling and measurement. Real loudspeakers approximate monopole behavior at low frequencies. They serve as fundamental building blocks in acoustics.

Muddy — A tonal descriptor indicating excessive low-mid buildup or lack of clarity. Muddy sound obscures detail and reduces intelligibility. It often results from room modes, poor EQ, or microphone placement. Corrective EQ and treatment restore definition.

Mult — A passive splitter that duplicates an audio signal to multiple destinations. Mults preserve signal integrity without active electronics. They are common in patch bays and modular systems. Proper use avoids loading or level issues.

Multi-Path — A condition where a signal arrives at a receiver via multiple paths due to reflections or obstacles. Multi-path causes interference, dropouts, or phase distortion in wireless systems. It is common in complex environments. Diversity reception helps mitigate multi-path issues.

Multitrack — A recording method that captures separate audio sources on individual tracks. Multitrack recording allows detailed editing, mixing, and processing. It revolutionized modern music production. Tracks can be layered, replaced, or rearranged freely.

Mute — A control that silences an audio channel without altering other settings. Muting helps manage routing, troubleshooting, and performance cues. It is essential in both live and studio workflows. Mute switches provide instant, reliable control.

Mute Group — A set of channels that can be muted or unmuted simultaneously. Mute groups streamline complex live sound operations. They allow quick transitions between scenes or ensembles. Proper grouping improves workflow efficiency.

Muting — The act of silencing a channel or system using a mute control. Muting prevents unwanted noise, feedback, or bleed. It is used during transitions, troubleshooting, and cueing. Effective muting supports clean, controlled audio.

NAB (National Association of Broadcasters) — A U.S. trade association representing radio and television broadcasters, known for developing influential technical standards and operational practices. In audio, NAB standards historically defined tape equalization curves, recording levels, and mechanical specifications for professional analog tape machines. The organization also shapes policies, engineering guidelines, and broadcast-quality requirements that affect signal processing, loudness management, and transmission workflows. NAB's conventions and publications continue to guide best practices across the broadcast industry.

NAG (Needed Acoustic Gain) — The amount of gain a sound-reinforcement system must provide so that listeners hear the source at the desired level. NAG depends on factors such as the distance between the talker and microphone, the distance between the loudspeaker and audience, and the target loudness for intelligibility. Comparing NAG with PAG (Potential Acoustic Gain) reveals whether a system can achieve the required loudness without feedback. This relationship is central to designing stable, intelligible reinforcement for speech and live sound.

Narrow-Band Noise — Noise concentrated within a small frequency range rather than spread broadly across the spectrum. It is useful for diagnostic testing, masking studies, and evaluating system response. Narrow-band noise reveals resonances and modal behavior more clearly than broadband signals. It is widely used in acoustics research and measurement.

Natural Quiet — An environment where human-made noise is minimal or absent, allowing natural soundscapes to dominate. Natural quiet is valued in wilderness areas, parks, and conservation zones. It supports wildlife behavior, visitor experience, and ecological balance. Preserving natural quiet requires careful noise management.

Near Field — The region close to a sound source where sound pressure and particle velocity behave irregularly. In the near field, directional patterns and level decay differ from far-field behavior. Measurements taken here require specialized techniques. Many microphones and loudspeakers exhibit unique characteristics in the near field.

Near-Field Scanning — A measurement technique that maps sound pressure close to a source to reconstruct its far-field behavior. It captures detailed spatial information for loudspeaker design, diagnostics, and quality control. Near-field scanning reveals complex radiation patterns. Advanced algorithms convert near-field data into accurate far-field predictions.

Newtonian Speed of Sound — The theoretical speed of sound derived under the assumption of isothermal conditions. Newton's model underestimates real sound speed because air behaves adiabatically during rapid compression. Later corrections by Laplace improved accuracy. The concept illustrates the evolution of acoustic theory.

NODSS (National Oceanic Data Storage System) — A digital data-management and archiving framework used in oceanographic and underwater-acoustics research to store, organize, and retrieve large volumes of observational data. NODSS supports long-term preservation of hydroacoustic recordings, environmental measurements, and sensor outputs collected by research vessels, buoys, and monitoring networks. Its standardized formats and metadata practices ensure that acoustic datasets remain accessible, comparable, and usable for scientific analysis across institutions.

Noise — Unwanted or undesirable sound that interferes with communication, comfort, or perception. Noise may be continuous, intermittent, or impulsive. Its impact depends on level, frequency, and context. Managing noise is central to acoustics, engineering, and environmental design.

Noise Criteria — A set of standardized curves used to evaluate background noise levels in indoor spaces. Noise Criteria (NC) ratings help assess comfort, speech intelligibility, and HVAC performance. Lower NC values indicate quieter environments. They are widely used in building design and commissioning.

Noise Criterion Curves — Graphs that define acceptable background noise levels across frequency bands for various room types. Designers compare measured spectra to these curves to determine compliance. Proper alignment ensures comfort and functional performance. NC curves guide HVAC and architectural decisions.

Noise Emission — The sound produced by a device, machine, or system during operation. Noise emission levels are measured for regulatory compliance, product labeling, and environmental impact. Lower emissions improve comfort and safety. Manufacturers work to reduce emission through design and materials.

Noise Floor — The baseline level of noise present in a system or environment when no intentional signal is present. A low noise floor allows greater dynamic range and cleaner audio. It is influenced by electronics, room acoustics, and environmental factors. Engineers strive to minimize noise floor for high-quality recordings.

Noise Immission — The noise that actually reaches a listener or receiver after propagation and environmental effects. Immission differs from emission because it accounts for distance, barriers, and reflections. It is central to environmental noise assessment. Regulations often specify immission limits.

Noise Induced Permanent Threshold — A long-term elevation in hearing threshold caused by repeated or intense noise exposure. This permanent shift indicates irreversible hearing damage. It affects communication, safety, and quality of life. Prevention relies on protection and exposure control.

Noise Induced Temporary Threshold — A short-term elevation in hearing threshold following noise exposure. Hearing typically recovers over time, but repeated temporary shifts can lead to permanent damage. Temporary threshold shifts indicate excessive exposure. Monitoring them helps prevent long-term harm.

Noise Isolation Class — A rating that measures the ability of a building partition to block airborne sound. NIC values are based on field measurements rather than laboratory tests. Higher NIC ratings indicate better isolation. NIC helps evaluate real-world acoustic performance.

Noise Mapping — The process of creating visual maps that show noise levels across geographic areas. Noise maps support urban planning, transportation design, and environmental regulation. They reveal hotspots, trends, and exposure patterns. Advanced models incorporate traffic, terrain, and meteorology.

Noise Power Emission Level — A metric describing the total acoustic power emitted by a device. It provides a standardized way to compare noise output across products. Manufacturers use it for labeling, compliance, and design optimization. Lower emission levels improve user comfort.

Noise Rating Curves — A set of curves used to evaluate background noise in buildings, similar to NC curves but with different criteria. Noise Rating (NR) curves are widely used internationally. They help ensure acceptable acoustic conditions for various room types. NR ratings guide HVAC and architectural design.

Noise Reduction — The decrease in sound level achieved by barriers, treatments, or system design. Noise reduction improves comfort, privacy, and communication. It may involve absorption, isolation, or electronic processing. Effective noise reduction requires understanding source, path, and receiver.

Noise Reduction Between Rooms — The decrease in sound level measured between two adjacent spaces. It depends on wall construction, sealing, and flanking paths. High reduction improves privacy and comfort. Field measurements often differ from laboratory predictions.

Noise Reduction Coefficient — A single-number rating representing the average absorption of a material across key mid-frequency bands. NRC values range from 0 (no absorption) to 1 (full absorption). Higher NRC indicates better sound absorption. It is widely used in selecting acoustic treatments.

Nom (Nominal Level) — A reference signal level used to describe the intended or standard operating point of an audio system. Nominal level indicates the typical voltage or digital value at which equipment is designed to run, leaving headroom above it for peaks and ensuring low noise below it. In analog systems, nominal levels are often expressed in dBu or dBV; in digital systems, they relate to a level below full scale (such as −20 dBFS). Using a consistent nominal level helps maintain proper gain structure, minimize distortion, and ensure predictable performance across interconnected devices.

Non Diversity — A wireless system configuration that uses a single antenna rather than diversity reception. Non-diversity systems are more susceptible to dropouts and multi-path interference. They are suitable for simple, low-risk applications. Professional systems typically use diversity for reliability.

Nonlinear — Describing a system whose output is not directly proportional to its input. Nonlinear behavior introduces distortion, harmonics, or compression. Some nonlinearities are undesirable, while others are used creatively. Understanding nonlinearity is essential in audio design.

Nonlinear Acoustics — The study of sound behavior at high amplitudes where linear assumptions break down. Nonlinear effects include shock formation, harmonic generation, and waveform distortion. These phenomena occur in ultrasonics, explosions, and high-intensity sound fields. Nonlinear acoustics requires advanced modeling.

Non-Steady Noise — Noise that varies significantly over time in level or spectral content. Examples include traffic bursts, machinery cycles, and intermittent events. Non-steady noise requires specialized metrics for accurate assessment. It often has greater annoyance potential than steady noise.

Normal Incidence Sound Absorption — The absorption of sound striking a surface at a perpendicular angle. It is measured in impedance tubes and provides precise material data. Normal incidence values differ from random incidence absorption. These measurements support material design and modeling.

Normal Mode — A resonant pattern in a room or structure where standing waves form at specific frequencies. Normal modes create peaks and nulls in frequency response. They are most prominent at low frequencies. Managing modes improves clarity and tonal balance.

Normalized Noise Isolation Class — A rating that adjusts field-measured noise isolation to a standard reference condition. It allows comparison across different environments. NNIC accounts for room absorption and geometry. It is used in building acoustics evaluation.

Normalized Noise Reduction — A measure of noise reduction adjusted for room absorption and other variables. It provides a standardized comparison of partition performance. Normalized values help evaluate materials and construction methods. They support accurate acoustic design.

Normalled — A patch-bay configuration where connections are automatically routed until a plug is inserted to break the path. Normalled circuits streamline workflow and reduce cabling. They support flexible routing in studios and live systems. Understanding normalled behavior prevents signal confusion.

Notch Filter — A filter that sharply attenuates a narrow frequency band while leaving others largely unaffected. Notch filters remove hum, feedback frequencies, or tonal artifacts. They are used in system tuning, restoration, and measurement. Their precision makes them powerful corrective tools.

Null — A point or direction where sound level is significantly reduced due to cancellation or directional behavior. Nulls occur in polar patterns, room modes, and interference patterns. They help control bleed and feedback. Understanding nulls guides placement and tuning.

Nyquist Frequency — Half the sampling rate in digital audio, representing the highest frequency that can be accurately captured. Frequencies above Nyquist cause aliasing unless filtered. Proper anti-aliasing ensures clean digital recording. The Nyquist concept is fundamental to DSP.

Oblique Mode — A room mode involving reflections from all three pairs of parallel surfaces. Oblique modes are more numerous and complex than axial or tangential modes. They contribute to overall low-frequency behavior. Their distribution affects modal density and smoothness.

Octave — A frequency interval where the upper frequency is twice the lower frequency. Octaves reflect how humans perceive pitch and tonal spacing. Acoustic measurements often use octave or one-third-octave bands. Octave analysis simplifies complex spectra.

Octave Band — A frequency band whose upper limit is twice its lower limit. Octave bands align with how humans perceive pitch and are widely used in acoustic measurement. They simplify complex spectra into manageable segments. Octave-band analysis supports noise control, room design, and system tuning.

Octave Band Sound Pressure Level — The sound pressure level measured within a single octave band. It reveals how energy is distributed across broad frequency regions. This metric helps diagnose tonal imbalances and evaluate noise sources. It is fundamental in environmental and architectural acoustics.

Octave Bands — A set of standardized frequency bands, each spanning one octave. They provide a structured way to analyze and compare sound spectra. Octave bands are used in noise regulation, product testing, and room assessment. Their simplicity makes them a universal tool in acoustics.

Ohm — The unit of electrical resistance, symbolized by Ω. In audio, ohms describe the impedance of loudspeakers, headphones, and circuits. Proper matching ensures efficient power transfer and prevents overload. Ohmic values guide system design and compatibility.

Omnidirectional — Describing a microphone or loudspeaker pattern that radiates or receives sound equally in all directions. Omnidirectional devices capture natural ambience and consistent tonal balance. They are ideal for room measurements and certain recording techniques. Their lack of rejection requires careful placement.

On-Board Device — A component or processor built directly into a piece of equipment rather than added externally. On-board devices include EQs, effects, preamps, and DSP modules. They streamline workflow and reduce cabling. Their performance varies with design quality.

Onethird Octave Bands — Frequency bands that divide each octave into three narrower segments. One-third-octave analysis provides finer spectral detail than full-octave bands. It is used in noise control, product testing, and room tuning. These bands align closely with human auditory filters.

Open — A term describing a circuit or connection that is not electrically complete. An open path prevents current flow and may cause silence or malfunction in audio systems. Opens can result from broken cables, switches, or connectors. Troubleshooting restores proper continuity.

Open Circuit Voltage — The voltage present at a device's output when no load is connected. It represents the maximum potential difference available. In microphones, open-circuit voltage helps define sensitivity. Accurate measurement requires high-impedance instruments.

Optical Cable — A cable that transmits digital audio or data using light rather than electrical signals. Optical cables offer immunity to electromagnetic interference and support long distances. Common formats include TOSLINK and ADAT. They provide clean, reliable digital connectivity.

Oscillation — A repetitive variation in a signal, often caused by feedback or instability. Unwanted oscillation produces tones or noise that disrupt audio systems. Controlled oscillation forms the basis of synthesis and test signals. Stability design prevents unintended oscillation.

Oscillator — A device that generates periodic waveforms such as sine, square, or sawtooth waves. Oscillators are used for testing, synthesis, and modulation. Their frequency stability and purity affect measurement accuracy. They are fundamental tools in audio engineering.

Oscilloscope — An instrument that displays electrical waveforms over time. Oscilloscopes reveal amplitude, frequency, distortion, and transient behavior. They are essential for diagnosing audio circuits and verifying signal integrity. Modern digital scopes offer advanced analysis features.

Ossicles — The three tiny bones in the middle ear: malleus, incus, and stapes. They amplify and transmit sound vibrations from the eardrum to the cochlea. Their mechanical leverage enhances hearing sensitivity. Damage to the ossicles impairs sound transmission.

Out of Phase — A condition where two signals have mismatched timing or polarity, causing partial or complete cancellation. Out-of-phase signals weaken bass, alter imaging, and reduce clarity. Phase alignment is critical in multi-mic setups and loudspeaker arrays. Correct polarity ensures coherent sound.

Outboard Device — External audio equipment used alongside a mixer or interface. Outboard devices include compressors, EQs, reverbs, and effects processors. They offer specialized control and sonic character. Proper routing integrates them cleanly into the signal chain.

Outdoor–Indoor Transmission Class — A rating that measures how well a building façade reduces outdoor noise entering indoor spaces. OITC emphasizes low-frequency performance typical of traffic and aircraft noise. Higher values indicate better isolation. It is widely used in urban and transportation planning.

Output — The signal or power delivered by a device to another component or system. Output level determines compatibility, headroom, and noise performance. Outputs may be analog, digital, balanced, or unbalanced. Proper matching ensures clean signal transfer.

Output Power — The electrical or acoustic power delivered by an amplifier or loudspeaker. Higher output power supports greater loudness and dynamic capability. It must be matched to speaker ratings to avoid damage. Power specifications guide system design.

Oval Window — A membrane-covered opening in the inner ear where the stapes transmits vibrations into the cochlea. It converts mechanical motion into fluid waves. Proper function is essential for hearing sensitivity. Damage or stiffness reduces auditory efficiency.

Overload — A condition where a device receives more signal or power than it can handle. Overload causes distortion, clipping, or equipment damage. Indicators and limiters help prevent overload. Proper gain staging maintains safe operating levels.

Oversampling — A digital processing technique that samples audio at a multiple of the base rate. Oversampling reduces aliasing, improves filter performance, and enhances fidelity. It is common in converters, synthesis, and DSP algorithms. Higher oversampling yields cleaner results.

Overtone — A frequency component above the fundamental frequency. Overtones include both harmonics and non-harmonic partials. They shape timbre and tonal richness. Their distribution defines the character of instruments and voices.

Pad — An attenuator that reduces signal level before processing. Pads prevent overload in microphones, preamps, and mixers. They are essential when capturing loud sources. Engaging a pad preserves clarity and headroom.

PAG (Potential Acoustic Gain) — A predictive measure used in sound-reinforcement design to estimate how much gain a system can provide before feedback occurs. PAG is calculated from factors such as microphone-to-source distance, loudspeaker-to-listener distance, room acoustics, and system geometry. Comparing PAG with NAG (Needed Acoustic Gain) helps engineers determine whether a reinforcement system can meet intelligibility and loudness requirements without instability. PAG is central to designing stable microphone–loudspeaker layouts in live-sound and installed-sound environments.

Pan — The distribution of a signal across the stereo field. Panning creates spatial placement and separation between elements. It shapes imaging, depth, and mix clarity. Effective panning enhances listener engagement.

Parabolic Equation Method — A computational technique for modeling long-range sound propagation in complex environments. It approximates wave behavior efficiently over large distances. The method is used in environmental acoustics, underwater acoustics, and atmospheric modeling. It handles refraction, terrain effects, and atmospheric gradients.

Parabolic Microphone — A microphone system that uses a parabolic reflector to focus sound onto a small pickup element. It provides extreme directionality and long-distance capture. Parabolic mics are used in sports, wildlife recording, and surveillance. Their narrow focus enhances distant detail.

Parametric Array — A highly directional sound beam created using ultrasonic carriers and nonlinear interaction in air. Parametric arrays produce narrow, focused audio that can be aimed precisely. They are used in museums, signage, and specialized communication. Their behavior relies on nonlinear acoustics.

Parametric Equalizer — An equalizer that allows control over frequency, bandwidth (Q), and gain for each band. Parametric EQs offer precise tonal shaping and problem correction. They are essential in mixing, mastering, and system tuning. Their flexibility surpasses fixed-band EQs.

Partial — Any frequency component of a sound, including the fundamental and its overtones. Partials define timbre and harmonic structure. Not all partials are harmonically related. Their distribution varies across instruments and voices.

Particle Velocity — The speed at which air particles oscillate as a sound wave passes. Particle velocity is highest near sources and boundaries. It influences absorption, transducer behavior, and near-field effects. Measuring it provides insight beyond sound pressure alone.

Passive Absorber — An acoustic treatment that absorbs sound without electronic components. Passive absorbers include porous panels, bass traps, and resonators. They reduce reflections, reverberation, and modal buildup. Their performance depends on material, thickness, and placement.

Passive Circuitry — Electronic circuits that operate without external power, using only resistors, capacitors, and inductors. Passive circuits shape tone, filter signals, and divide frequencies. They are reliable and noise-free but limited in gain. Many crossovers and EQ networks are passive.

Passive Crossover — A crossover network placed after the amplifier to divide frequencies between loudspeaker drivers. Passive crossovers use inductors, capacitors, and resistors. They simplify wiring but introduce power loss and limited flexibility. Proper design ensures smooth transitions between drivers.

Patch — A connection made using cables or a patch bay to route audio signals between devices. Patching enables flexible signal flow in studios and live systems. It supports creative routing, troubleshooting, and system organization. Clear labeling prevents confusion.

Patch Bay — A centralized panel that allows flexible routing of audio signals using patch cables. Patch bays organize complex studio or live-sound setups by bringing hidden connections to an accessible interface. They support quick reconfiguration without rewiring equipment. Clear labeling ensures efficient workflow and prevents signal confusion.

Patch Panel — A structured panel used to terminate and route audio, data, or power connections. Patch panels are common in broadcast, IT, and installed-sound environments. They provide organized, serviceable access to wiring infrastructure. Proper layout improves reliability and maintenance.

Patch Point — A connection location on a mixer or device where a signal can be accessed or redirected. Patch points support inserts, sends, and external processing. They enable flexible routing within a signal chain. Understanding patch points is essential for advanced system configuration.

Peak — The highest instantaneous value of a waveform. Peaks determine headroom requirements and influence distortion risk. Peak measurement is essential in dynamics processing and system protection. Managing peaks ensures clean, undistorted audio.

Peak A-Weighted Sound Pressure Level — The maximum instantaneous sound pressure level measured using the A-weighting curve. It reflects perceived loudness while emphasizing mid-frequency sensitivity. This metric is used in noise regulation and hearing-risk assessment. It captures extreme events more realistically than unweighted peaks.

Peak Light — A visual indicator on audio equipment that signals when a signal is approaching or exceeding safe operating levels. Peak lights help prevent clipping and distortion. They provide quick feedback during recording and mixing. Proper monitoring maintains clean gain structure.

Peak Limiter — A dynamics processor designed to prevent signal peaks from exceeding a set threshold. Limiters protect equipment, maintain broadcast compliance, and control transient spikes. They act quickly to catch fast peaks. Proper settings preserve loudness while avoiding audible artifacts.

Peak Sound Pressure — The maximum instantaneous pressure variation in a sound wave. It represents the extreme amplitude of a signal. Peak sound pressure is important in evaluating impulsive noise and transducer performance. It differs from RMS or average measures.

Peak Sound Pressure Level — The decibel value corresponding to the highest instantaneous sound pressure during a measurement. It is used for assessing impact noise, explosions, and other transient events. Peak SPL helps determine hearing-protection requirements. It captures short-duration extremes that average metrics miss.

Peaking — A type of equalization that boosts or cuts a specific center frequency with adjustable bandwidth. Peaking filters shape tonal character and correct resonances. They are common in parametric EQs. Proper use enhances clarity and balance.

Peak-To-Peak Value — The total amplitude difference between the highest positive and lowest negative points of a waveform. It provides a full picture of signal excursion. Peak-to-peak values are used in electronics, acoustics, and measurement. They help evaluate headroom and distortion potential.

Peclet Number — A dimensionless number describing the relative importance of convection to diffusion in fluid flow. In acoustics, it appears in modeling heat transfer and flow-related noise. High Peclet numbers indicate convection-dominated behavior. It is relevant in aeroacoustics and environmental modeling.

Perceived Noise Level — A psychoacoustic metric that estimates how noisy a sound is to human listeners. It accounts for frequency content, tonal components, and annoyance factors. Perceived noise level is used in aviation, transportation, and product design. It provides a more realistic assessment than simple SPL.

Percentile — A statistical measure indicating the value below which a certain percentage of observations fall. In acoustics, percentile levels describe fluctuating noise, such as L10 or L90. They reveal patterns of peaks and background levels. Percentiles support environmental and occupational noise analysis.

Period — The time required for one complete cycle of a periodic waveform. Period is the inverse of frequency. It determines timing, pitch, and waveform structure. Understanding period is fundamental to signal analysis.

Periodic Quantity — A physical quantity that repeats at regular intervals. Periodic signals include tones, vibrations, and oscillations. Their behavior is defined by frequency, amplitude, and phase. Periodicity underlies harmonic structure and spectral analysis.

Permanent Threshold Shift — A lasting increase in hearing threshold caused by prolonged or intense noise exposure. Unlike temporary shifts, permanent threshold shifts do not recover. They indicate irreversible damage to hair cells or auditory pathways. Prevention relies on proper hearing protection and exposure control.

PFC (Power Factor Correction) — A technique used in audio power supplies and electrical systems to improve the alignment between voltage and current, reducing reactive power and increasing overall efficiency. In professional audio, PFC helps amplifiers and digital processors draw power more cleanly, lowering harmonic distortion on the mains and minimizing stress on circuits. Systems with active PFC maintain stable operation under varying loads, support higher peak demands, and reduce the likelihood of noise or interference caused by poor power quality.

PFL (Pre-Fader Listen) — A solo-monitoring mode on a mixing console that lets the engineer hear a channel *before* its fader and most processing are applied. PFL routes the raw input signal to the control-room or headphone bus without affecting the main mix, making it ideal for setting input gain, checking signal integrity, and diagnosing noise or distortion. Because it bypasses the channel fader, PFL provides a consistent reference level for accurate gain staging.

Phantom Power — A DC voltage, typically +48 V, supplied through microphone cables to power condenser microphones and active DI boxes. Phantom power travels on balanced lines without affecting audio. It enables compact, reliable microphone operation. Proper use avoids damage to incompatible equipment.

Phase — The relative timing relationship between periodic waveforms. Phase affects reinforcement, cancellation, and spatial imaging. Misaligned phase can cause comb filtering or tonal changes. Managing phase is essential in multi-mic and multi-speaker setups.

Phase Shift — A change in the phase relationship between two signals. Phase shifts occur in filters, delays, and propagation paths. They influence tonal balance and interference patterns. Understanding phase shift helps maintain coherent sound.

Phase Velocity — The speed at which a particular phase point of a wave travels through a medium. Phase velocity may differ from group velocity in dispersive media. It influences propagation, refraction, and wave behavior. It is important in acoustics, optics, and materials science.

Phasing — A modulation effect created by mixing a signal with a slightly delayed, frequency-swept version of itself. Phasing produces sweeping notches and a swirling character. It is used creatively in music production. The effect arises from comb-filter interactions.

Phon — A unit of perceived loudness based on equal-loudness contours. A sound has a loudness of X phons if it is perceived as loud as a pure tone at X dB SPL at 1 kHz. Phons reflect human hearing sensitivity across frequencies. They support psychoacoustic analysis.

Phone Plug — A common audio connector, typically referring to 1/4-inch TS or TRS plugs. Phone plugs carry instrument, line, or headphone signals. They are durable and widely used in professional audio. Proper wiring ensures noise-free operation.

Phones — A shorthand term for headphones or headphone outputs. "Phones" jacks provide amplified signals suitable for personal monitoring. They support mixing, tracking, and casual listening. Their quality affects clarity and comfort.

Phono Plug — A connector used for turntable inputs requiring special equalization and grounding. Phono plugs typically refer to RCA connectors used in consumer audio. They carry low-level signals that require careful shielding. Proper phono connections preserve fidelity and reduce hum.

Physical Meteorology — The study of atmospheric physical processes such as temperature, humidity, and wind. These factors influence outdoor sound propagation, refraction, and absorption. Physical meteorology is essential in environmental acoustics and noise prediction. Weather conditions can dramatically alter sound behavior.

Pick-Up — The process by which a microphone or transducer captures sound. Pick-up quality depends on pattern, placement, and environment. Good pick-up ensures clarity and isolation. Poor pick-up introduces noise, bleed, or coloration.

Pick-Up Angle — The angular region in which a microphone effectively captures sound. It corresponds to the main lobe of the polar pattern. Wider angles capture more ambience; narrower angles improve isolation. Understanding pick-up angle guides microphone placement.

Piezoelectric Ceramic — A material that generates electrical charge when mechanically stressed. Piezoelectric ceramics are used in sensors, buzzers, and audio transducers. Their high stiffness and sensitivity make them ideal for high-frequency applications. They form the basis of many compact audio devices.

Piezoelectric Microphone — A microphone that uses piezoelectric materials to convert vibration into electrical signals. These microphones are rugged, inexpensive, and suitable for contact or instrument pickup. They excel at capturing mechanical vibrations. Their frequency response differs from traditional air-coupled microphones.

Piezoelectric Tweeter — A high-frequency loudspeaker that uses piezoelectric materials to produce sound. Piezo tweeters are efficient, lightweight, and require no crossover. They are common in portable and budget systems. Their tonal character differs from dynamic or ribbon tweeters.

Pink Noise — A noise signal with equal energy per octave, resulting in a natural, balanced spectral slope. Pink noise is used for system tuning, measurement, and psychoacoustic testing. It resembles many real-world sound distributions. Its spectral shape makes it ideal for calibration.

Pinna — The outer, visible part of the ear that collects and shapes incoming sound. The pinna introduces direction-dependent filtering essential for localization. Its folds and geometry help distinguish elevation and front-back cues. It plays a crucial role in spatial hearing.

Pitch — The perceived highness or lowness of a sound, determined primarily by its fundamental frequency. Pitch perception is influenced by harmonics, loudness, and auditory context. It is central to music, speech, and psychoacoustics. Accurate pitch reproduction depends on stable frequency response.

Place Effect — A psychoacoustic phenomenon where pitch perception is influenced by the location of maximum vibration along the basilar membrane. Higher frequencies stimulate regions near the base of the cochlea, while lower frequencies stimulate regions near the apex. The place effect supports frequency discrimination. It is fundamental to auditory physiology and hearing models.

Planar Transducer — A loudspeaker or headphone driver that uses a flat diaphragm driven across its entire surface. Planar transducers offer low distortion, fast response, and detailed sound. They rely on magnetic or electrostatic forces distributed evenly. Their design provides excellent clarity and transient accuracy.

Plane Wave — A sound wave whose wavefronts are parallel and propagate in a uniform direction. Plane waves approximate sound behavior far from a source or in controlled environments. They simplify acoustic modeling and measurement. Real-world sound fields often combine plane and spherical components.

Player — A device or software application that plays back audio content. Players range from portable devices to professional media servers. Their quality affects fidelity, stability, and user experience. Modern players support multiple formats and streaming protocols.

Plenum — An air-handling space used for HVAC distribution, often above ceilings or below floors. Plenums can transmit noise if not properly treated. Acoustic lining and isolation reduce airflow noise and mechanical transmission. Plenum design influences building acoustics.

Plug — A connector used to terminate or interface audio cables. Plugs include TS, TRS, XLR, and RCA types. Their quality affects reliability, noise rejection, and durability. Proper mating ensures clean signal transfer.

Point Source — An idealized sound source that radiates uniformly in all directions from a single point. Real loudspeakers approximate point-source behavior at certain frequencies. Point-source models simplify prediction of level decay and coverage. They are foundational in acoustic theory.

Polar Pattern — A graph showing a microphone or loudspeaker's sensitivity as a function of angle. Patterns include cardioid, omnidirectional, and figure-eight. Polar patterns determine isolation, feedback resistance, and tonal consistency. Understanding them guides placement and system design.

Polarity — The orientation of a signal's positive and negative phases. Correct polarity ensures coherent summation between microphones and speakers. Reversed polarity causes cancellation, weak bass, and imaging issues. Polarity checks are essential in system alignment.

Pop — A short, low-frequency burst caused by plosive consonants or sudden pressure changes near a microphone. Pops distort recordings and reduce intelligibility. Proper mic technique and filtering prevent them. Pop control is essential in vocal production.

Pop Filter — A screen placed in front of a microphone to reduce plosive bursts. Pop filters diffuse airflow without affecting tonal balance. They are standard in vocal recording. Proper placement improves clarity and protects the microphone.

Porous Sound Absorber — An absorber made of fibrous or open-cell materials that dissipate sound through friction. Porous absorbers are effective at mid- and high-frequencies. Their performance depends on thickness, density, and airflow resistance. They are widely used in studios and architectural acoustics.

Post-Fader — A signal routing configuration where the send level follows the channel fader position. Post-fader sends are used for effects that should track mix balance. They ensure consistent wet/dry relationships. This routing is common for reverbs and delays.

Pot — A shorthand term for potentiometer, referring to a variable control knob. Pots adjust level, tone, or other parameters in audio equipment. Their taper and range influence user control. High-quality pots provide smooth, noise-free operation.

Potentiometer — A variable resistor used to control voltage or signal level. Potentiometers appear in volume controls, EQ circuits, and mixers. Their design affects precision, noise, and durability. They are fundamental components in analog audio systems.

Power Compression — A reduction in loudspeaker output caused by heating of the voice coil during high-power operation. As resistance increases, efficiency drops and output decreases. Power compression limits dynamic performance. Proper cooling and design minimize the effect.

Power Conditioner — A device that stabilizes and filters AC power to protect audio equipment. Conditioners reduce noise, surges, and voltage fluctuations. They improve reliability in studios and live environments. Clean power supports consistent performance.

Power Sequencer — A device that turns audio equipment on and off in a controlled order. Sequencing prevents pops, surges, and overloads during power transitions. It is essential in large systems with multiple amplifiers and processors. Proper sequencing protects equipment and speakers.

Power Spectrum Density — A measure of how signal power is distributed across frequency. PSD reveals tonal content, noise characteristics, and system behavior. It is used in analysis, modeling, and diagnostics. PSD plots support detailed spectral evaluation.

PRD (Primitive Root Diffuser) — A type of acoustic diffuser whose surface pattern is generated from a mathematical sequence based on primitive roots modulo a prime number. PRDs scatter sound energy broadly and evenly by using wells of varying depths arranged in a non-repeating pattern. Compared with quadratic residue diffusers, PRDs can offer smoother diffusion over wider bandwidths and greater design flexibility. They are used in studios, performance spaces, and control rooms to reduce specular reflections while preserving a lively, natural acoustic field.

Preamplifier — A device that boosts low-level signals, such as those from microphones or instruments, to line level. Preamps influence noise, clarity, and tonal character. High-quality preamps provide clean gain with minimal coloration. They are essential in recording and reinforcement.

Precedence Effect — A psychoacoustic phenomenon where the first arriving sound dominates localization, even when later reflections are louder. This effect helps listeners localize sources in reverberant spaces. It guides loudspeaker placement and room design. Proper control enhances clarity and imaging.

Pre-Emphasis — A filtering process that boosts high frequencies before transmission or recording. It improves signal-to-noise ratio by reducing the relative impact of noise. A matching de-emphasis filter restores natural balance on playback. Pre-emphasis is used in FM broadcasting and some digital formats.

Pre-Fader — A routing configuration where the send level is independent of the channel fader. Pre-fader sends are used for monitor mixes that must remain stable regardless of mix changes.

They provide performers with consistent cues. This routing is essential in live sound.

Presence — A tonal quality associated with clarity and forwardness in the upper midrange. Presence enhances speech intelligibility and vocal definition. Excessive presence can sound harsh, while too little sounds dull. Presence shaping is common in EQ and microphone design.

Presence Peak — A rise in frequency response in the upper midrange, often around 3–6 kHz. Many microphones include a presence peak to enhance articulation and clarity. It improves intelligibility in vocals and speech. The peak's shape influences tonal character.

Pressure Zone — A region near a boundary where sound pressure is reinforced due to reflection. Pressure zones provide strong, consistent signals with reduced phase interference. They are used in specialized microphone designs. Boundary effects shape low-frequency behavior.

Pressure Zone Microphone — A microphone that uses a boundary surface to capture sound in the pressure zone. PZMs offer wide coverage, strong clarity, and reduced phase issues. They are used in conference rooms, theaters, and ambient recording. Their design enhances intelligibility and spatial consistency.

Processor — A device or software tool that modifies audio through EQ, dynamics, effects, or other transformations. Processors shape tone, control levels, and enhance clarity. They are essential in mixing, mastering, and system tuning. Digital processors offer extensive flexibility.

Program — The audio content being played, mixed, or transmitted. Program material includes music, speech, effects, and broadcast content. Its characteristics influence system design and tuning. Program level and dynamics guide processing decisions.

Projector — A loudspeaker designed to project sound over long distances with controlled directivity. Projectors are used in paging, outdoor systems, and large venues. Their horn-loaded designs enhance efficiency and focus. Proper placement ensures intelligible coverage.

Proximity Effect — A bass boost that occurs when directional microphones are placed very close to a sound source. It enhances warmth and fullness but can cause muddiness if excessive. Singers and engineers use proximity effect creatively. Proper technique manages its impact.

Proximity Sensor — A device that detects the presence or movement of objects without physical contact. In audio, proximity sensors may trigger cues, automate controls, or support interactive installations. They use infrared, ultrasonic, or capacitive technologies. Their behavior depends on range and environment.

Psychoacoustic Annoyance — A metric that quantifies how irritating or unpleasant a sound is to listeners. It considers loudness, sharpness, roughness, and fluctuation strength. Psychoacoustic annoyance is used in product design, noise regulation, and environmental assessment. It reflects subjective response more accurately than SPL alone.

Psychoacoustics — The study of how humans perceive sound, including loudness, pitch, timbre, and spatial cues. Psychoacoustics bridges physics, physiology, and psychology. It informs audio coding, system design, and product evaluation. Understanding perception leads to more effective and engaging sound experiences.

Public Address System — A sound reinforcement system designed to amplify speech or music for an audience. PA systems include microphones, mixers, amplifiers, and loudspeakers arranged for clear, even coverage. They are used in venues ranging from classrooms to stadiums. Proper design ensures intelligibility, reliability, and appropriate loudness.

Pure Tone — A sound consisting of a single frequency with no harmonics or overtones. Pure tones are used in hearing tests, calibration, and psychoacoustic research. They produce a clean, sinusoidal waveform. Their simplicity makes them ideal for controlled measurement.

Q — A dimensionless parameter describing the sharpness or selectivity of a resonance or filter. Higher Q values indicate narrower bandwidth and greater precision. Q is central to EQ design, loudspeaker tuning, and modal analysis. It shapes how systems respond to specific frequencies.

Q-Factor — A measure of how sharply a filter or resonant system responds around its center frequency. High Q produces narrow, focused peaks or cuts; low Q yields broader adjustments. Q-factor influences tonal shaping, feedback control, and system stability. It is fundamental in filter design.

QRD — Quadratic Residue Diffuser, an acoustic device that scatters sound using mathematically derived well depths. QRDs reduce flutter echo and improve spatial uniformity without deadening a room. They are common in studios, halls, and listening rooms. Their performance depends on design sequence and depth.

Quantization Noise — The noise introduced when converting analog signals to digital form due to rounding errors. It appears as low-level hiss or grain in digital audio. Higher bit depth reduces quantization noise. Proper dithering minimizes artifacts.

Quasi-Impulsive Noise — Noise consisting of short bursts that are not fully impulsive but still exhibit rapid onset. It is common in mechanical systems, switching devices, and environmental sources. Quasi-impulsive noise requires specialized metrics for accurate assessment. It can be more disturbing than steady noise.

R — A context-dependent symbol that may represent resistance, reflectance, or room constant. Its meaning depends on the equation or acoustic model. Clear notation ensures proper interpretation. R is widely used across acoustics and electronics.

Rack — A standardized frame used to mount audio equipment in a vertical stack. Racks organize gear, improve airflow, and simplify cabling. They are common in studios, live sound, and broadcast facilities. Rack systems support efficient workflow and maintenance.

Rack-Mount — Describing equipment designed to fit into a standard 19-inch rack. Rack-mount gear includes processors, amplifiers, interfaces, and power units. It provides secure installation and consistent form factor. Rack-mounting improves organization and accessibility.

Radiant Energy — Energy transmitted through electromagnetic waves, such as light or infrared. In acoustics, radiant energy influences measurement devices and environmental conditions. It differs from acoustic energy, which travels through mechanical waves. Understanding radiant energy helps interpret sensor behavior.

Radio Frequency — The portion of the electromagnetic spectrum used for wireless communication. RF systems include wireless microphones, transmitters, and receivers. Proper frequency coordination prevents interference and dropouts. RF behavior is central to modern audio workflows.

Random Incidence — Sound arriving from all directions with equal probability. Random incidence conditions occur in diffuse fields and reverberant rooms. Measurements under random incidence differ from normal incidence values. They are essential for evaluating absorption and scattering.

Random Noise — Noise with unpredictable amplitude and frequency characteristics. It includes broadband signals such as white or pink noise. Random noise is used for testing, masking, and calibration. Its statistical properties support analysis and modeling.

Range — The distance over which a sound source, microphone, or wireless system operates effectively. Range depends on power, environment, and directivity. Exceeding range reduces clarity or causes dropouts. Proper planning ensures reliable performance.

Rarefaction — The low-pressure phase of a sound wave where air particles spread apart. Rarefaction alternates with compression to form longitudinal waves. It influences particle velocity and acoustic energy flow. Understanding rarefaction is fundamental to wave behavior.

RASTI — Rapid Speech Transmission Index, a simplified measure of speech intelligibility. RASTI evaluates clarity based on modulation transfer characteristics. It is used in room assessment, public address design, and code compliance. Higher values indicate better intelligibility.

Ray — A simplified representation of sound propagation used in geometrical acoustics. Rays model reflections, refractions, and paths through space. They are useful for large-scale prediction and room design. Ray models complement wave-based approaches.

Ray Tracing — A computational method that simulates sound propagation using rays. It predicts reflections, absorption, and coverage in complex spaces. Ray tracing supports architectural acoustics, system design, and noise control. It provides visual insight into sound paths.

Rayleigh Integral — A mathematical expression describing sound radiation from vibrating surfaces. It forms the basis for many acoustic radiation models. The integral helps predict directivity and pressure fields. It is widely used in transducer and loudspeaker design.

Rayleigh Wave — A surface wave that travels along the boundary of a solid medium. Rayleigh waves combine longitudinal and vertical shear motion. They are important in seismology, vibration analysis, and structural acoustics. Their slow decay makes them influential in surface behavior.

Rayleigh–Jeans Law — A classical formula describing the spectral distribution of blackbody radiation at long wavelengths. It predicts energy proportional to temperature and frequency squared. The law fails at high frequencies, leading to the ultraviolet catastrophe. Its limitations helped inspire quantum theory.

RC — A context-dependent abbreviation that may refer to "room constant," "resistance-capacitance," or "rating class." Its meaning depends on the acoustic or electrical context. RC values guide design and analysis. Clear notation avoids confusion.

RCA Phono Plug — A consumer-grade audio connector used for line-level and phono-level signals. RCA plugs are common in home audio, turntables, and video equipment. They are unbalanced and susceptible to noise over long runs. Proper shielding improves performance.

R-DAT — Rotary Digital Audio Tape, a digital recording format using rotating heads similar to video tape systems. R-DAT offered high-quality digital recording in compact form. It was used in studios, field recording, and archiving. The format is now largely obsolete but historically significant.

Reach — The effective distance over which a microphone or loudspeaker maintains clarity and usable level. Reach depends on directivity, sensitivity, and environment. Greater reach improves intelligibility in large spaces. Proper design maximizes reach without excessive level.

Reactance — The opposition to alternating current caused by capacitance or inductance. Reactance varies with frequency and influences filter behavior. In acoustics, it appears in impedance models of transducers and materials. Understanding reactance supports system tuning.

Reactive Absorber — An absorber that uses resonance rather than porosity to dissipate sound energy. Reactive absorbers target narrow frequency ranges, especially at low frequencies. Examples include Helmholtz resonators and membrane traps. They complement broadband absorbers in room design.

Reactive Silencer — A noise-control device that uses resonant chambers to reduce specific frequencies. Reactive silencers are effective for tonal or low-frequency noise. They are used in HVAC, engines, and industrial systems. Their performance depends on geometry and tuning.

Receiver — A device that accepts audio signals, either through wireless transmission or wired input. Receivers decode, amplify, or route signals for playback or processing. Wireless receivers require proper frequency coordination. Their quality affects reliability and clarity.

Receiving Room — The room on the receiving side of a sound transmission test. Measurements in the receiving room determine isolation performance. Its absorption and volume influence results. Standardized conditions ensure accurate comparison.

Recorder — A device that captures audio for storage and playback. Recorders range from handheld units to multitrack digital systems. Their fidelity depends on converters, media, and processing. Recording is central to production, archiving, and research.

Recording — The process of capturing sound for later reproduction or analysis. Recording involves microphones, preamps, converters, and storage media. Technique and environment strongly influence quality. Good recording preserves detail, dynamics, and tonal balance.

Recording Material — The medium used to store recorded audio, such as tape, vinyl, or digital memory. Each material has unique characteristics, durability, and noise behavior. Material choice affects fidelity and longevity. Modern systems rely primarily on digital storage.

Reference Sound Source — A calibrated loudspeaker used to generate known sound levels for measurement. It ensures consistency in room acoustics, isolation tests, and equipment calibration. Reference sources follow strict standards for output and spectrum. They provide reliable baselines for comparison.

Reference Volume — A standardized volume used in acoustic calculations or calibration. Reference volumes help normalize measurements across different spaces. They support modeling, absorption testing, and system evaluation. Consistent reference values ensure comparability.

Reflection — The return of sound waves after striking a surface. Reflections shape room acoustics by adding reverberation, altering clarity, and influencing spatial impression. Their strength and timing depend on surface material and geometry. Managing reflections is essential for intelligibility and tonal accuracy.

Reflection-Phase Grating — A structured surface that uses periodic variations in depth or spacing to control the phase of reflected sound. These gratings scatter energy in predictable ways, enhancing diffusion and reducing coloration. They are used in studios, halls, and acoustic treatments. Proper design tailors scattering to specific frequency ranges.

Refraction — The bending of sound waves as they pass through regions with different temperatures, densities, or wind gradients. Refraction alters propagation paths, sometimes carrying sound farther or shadowing certain areas. It is important in outdoor acoustics and environmental modeling. Atmospheric conditions strongly influence refraction behavior.

Residual Noise — The background noise remaining after a sound source has stopped. Residual noise includes HVAC hum, traffic, and electronic hiss. It limits dynamic range and affects measurement accuracy. Low residual noise is essential for critical listening and precise testing.

Resistance — The component of impedance that dissipates energy as heat rather than storing it. In acoustics, resistance appears in porous absorbers and transducer models. It influences damping, efficiency, and frequency response. Proper resistance improves stability and reduces resonance.

Resonance — The amplification of vibration or sound at specific frequencies where a system naturally oscillates. Resonance can enhance tone or cause unwanted peaks and coloration. It depends on mass, stiffness, and damping. Controlling resonance is crucial in room design and equipment engineering.

Resonant Frequency — The frequency at which a system vibrates most efficiently with minimal external force. At resonance, amplitude increases and energy storage peaks. Resonant frequencies shape tonal character and modal behavior. Identifying them helps diagnose acoustic problems.

Resonator Silencer — A noise-control device that uses tuned cavities to reduce specific frequencies. Resonator silencers are effective for tonal or low-frequency noise in ducts and machinery. Their performance depends on cavity geometry and tuning. They complement absorptive silencers in HVAC design.

Response — The behavior of a system when stimulated by sound or electrical input. Response includes frequency response, transient response, and dynamic behavior. Accurate response ensures faithful reproduction. Measuring response guides tuning and optimization.

Return — A signal path that brings processed audio back into a mixer or system. Returns are used for effects, submixes, and external processing. They allow blending of wet and dry signals. Proper routing maintains clarity and balance.

Reverb — The dense collection of reflections that persist after the direct sound stops. Reverb adds spaciousness, depth, and realism to audio. Excessive reverb reduces intelligibility, while controlled reverb enhances musicality. It is central to room acoustics and effects processing.

Reverberant Sound Field — A sound field dominated by reflections rather than direct sound. In this region, sound energy becomes diffuse and uniform. Reverberant fields influence tonal balance and loudness. They are important in large rooms and performance spaces.

Reverberation — The persistence of sound due to repeated reflections in an enclosed space. Reverberation enriches music but can obscure speech. Its character depends on room size, shape, and absorption. Controlling reverberation is essential for functional acoustics.

Reverberation Chamber — A highly reflective room designed to create a diffuse, long-lasting reverberant field. These chambers are used for measuring absorption, power output, and material properties. Their geometry minimizes standing waves. They provide standardized test conditions.

Reverberation Radius — The distance from a sound source where direct and reverberant sound levels are equal. Inside this radius, direct sound dominates; beyond it, reverberation prevails. The radius depends on room absorption and volume. It is key to microphone placement and system design.

Reverberation Room — A room designed to produce strong, uniform reverberation for acoustic testing. Reverberation rooms support measurements of absorption, scattering, and sound power. Their surfaces are highly reflective and irregular. They follow strict standards for accuracy.

Reverberation Time — The time required for sound to decay by 60 dB after the source stops. Reverberation time (RT60) defines room acoustics and influences clarity, warmth, and envelopment. Ideal RT varies by room function. It is a primary metric in architectural acoustics.

Reverse Polarity — A condition where the positive and negative connections of a signal are swapped. Reverse polarity causes phase inversion, weakening bass and altering imaging. It is a common troubleshooting issue. Correct polarity ensures coherent summation.

Reynolds Number — A dimensionless number describing the ratio of inertial to viscous forces in fluid flow. In acoustics, it helps predict turbulence and flow-induced noise. High Reynolds numbers indicate turbulent behavior. It is central to aeroacoustics and HVAC design.

RF — Radio frequency, the electromagnetic spectrum used for wireless communication. RF systems include wireless microphones, transmitters, and receivers. Proper RF management prevents interference and dropouts. RF behavior is essential in modern audio workflows.

RF Noise — Unwanted electromagnetic interference within the radio frequency spectrum. RF noise disrupts wireless audio, communication, and sensitive electronics. Shielding, filtering, and proper grounding reduce RF noise. It is a major concern in live and broadcast environments.

RFI — Radio frequency interference, caused by unwanted RF signals entering audio equipment. RFI produces buzz, hiss, or distortion. Balanced lines, shielding, and filtering help prevent it. RFI control is essential for clean audio.

RFZ — Reflection-Free Zone, an acoustic design strategy that eliminates early reflections around the listening position. RFZ rooms use angled surfaces and absorption to preserve imaging and clarity. They are common in control rooms and mastering suites. Proper geometry ensures effectiveness.

RG-58U — A type of coaxial cable commonly used for RF and communication applications. RG-58U offers moderate shielding and flexibility. It is used in wireless antenna systems and instrumentation. Proper termination ensures low loss and stable performance.

RG-59U — A coaxial cable used for video, RF, and certain audio applications. RG-59U has higher impedance and lower loss at video frequencies. It is common in broadcast and CCTV systems. Quality shielding reduces interference.

Ribbon Microphone — A microphone that uses a thin metal ribbon suspended in a magnetic field to generate voltage. Ribbon mics offer smooth, natural high-frequency response and figure-eight patterns. They are prized for warmth and detail. Their delicate construction requires careful handling.

Ring — A tonal artifact caused by resonances in drums, rooms, or equipment. Ringing adds unwanted sustain or metallic character. It is controlled through damping, EQ, or treatment. Identifying ring frequencies improves clarity.

Ringing — The prolonged oscillation of a system after excitation. Ringing occurs in filters, rooms, and mechanical structures. Excessive ringing reduces clarity and introduces coloration. Proper damping and design minimize it.

RMS — Root Mean Square, a mathematical measure representing the effective value of a varying signal. RMS correlates closely with perceived loudness and energy. It is used in metering, power ratings, and analysis. RMS values provide stable, meaningful measurements.

Rolloff — The gradual attenuation of frequencies beyond a filter's cutoff point. Rolloff rate determines how sharply a filter reduces unwanted content. It shapes tonal balance and system behavior. Steeper rolloff provides stronger isolation.

Room Absorption — The total sound-absorbing capability of a room's surfaces and contents. Room absorption influences reverberation time, clarity, and tonal balance. Increasing absorption reduces reflections and reverberation. It is central to acoustic design.

Room Constant — A parameter describing the effective absorption of a room, combining surface absorption and geometry. The room constant helps predict sound level distribution and reverberation. Higher values indicate more absorptive spaces. It is used in Sabine-based calculations.

Room Criteria Curves — A set of curves used to evaluate background noise in buildings, focusing on HVAC and mechanical noise. RC curves emphasize low-frequency rumble and tonal components. They guide design for comfort and speech clarity. RC ratings complement NC and NR systems.

Room Mode — A standing wave pattern formed by reflections between room boundaries. Modes cause peaks and nulls in low-frequency response. Their frequencies depend on room dimensions. Managing modes improves bass accuracy and clarity.

Rotary Control — A knob or dial used to adjust audio parameters such as gain, EQ, or pan. Rotary controls offer smooth, continuous adjustment. Their feel and taper influence usability. They are ubiquitous in analog and digital equipment.

Roughness — A psychoacoustic attribute describing rapid amplitude fluctuations that create a harsh or coarse sound quality. Roughness peaks at modulation frequencies around 70 Hz. It influences perceived annoyance and sound quality. Designers use roughness metrics in product evaluation.

Round Window — A flexible membrane in the inner ear that allows fluid movement within the cochlea. It compensates for pressure changes created by the stapes at the oval window. Proper round-window function enables accurate transmission of sound waves. Damage or stiffness impairs hearing sensitivity.

RPG Cloud — A suspended acoustic diffuser panel designed by RPG to scatter sound energy overhead. RPG Clouds reduce flutter echo and improve spatial uniformity without deadening the room. They are common in studios, auditoriums, and performance spaces. Their placement enhances clarity and envelopment.

RT60 — The time required for sound to decay by 60 dB after the source stops. RT60 defines a room's reverberation characteristics and strongly influences clarity and warmth. Ideal values vary by room function. RT60 is a primary metric in architectural acoustics.

RU — Rack Unit, a standardized height measurement for rack-mount equipment. One RU equals 1.75 inches. It ensures compatibility across audio, broadcast, and IT gear. Rack planning relies on RU allocation.

Rumble — Low-frequency noise caused by mechanical vibration, turntable resonance, or structural transmission. Rumble masks bass detail and reduces clarity. Filters and isolation help control it. It is especially problematic in analog playback systems.

Sabin — A unit of sound absorption equal to one square foot of perfectly absorbing surface. Sabins quantify how much sound energy a material removes from a room. They support reverberation time calculations. The metric is central to architectural acoustics.

Sabine — A unit or method derived from Wallace Sabine's foundational work in room acoustics. Sabine's formula relates reverberation time to room volume and absorption. It is widely used for early-stage acoustic design. The Sabine approach assumes diffuse sound fields.

Sabine Absorption — The total absorption in a room expressed in sabins. Sabine absorption determines reverberation time and acoustic balance. Increasing absorption reduces reflections and reverberation. It is calculated from material coefficients and surface areas.

Sample Rate — The number of digital samples taken per second when converting analog audio to digital form. Higher sample rates capture more detail and reduce aliasing. Common rates include 44.1, 48, and 96 kHz. Sample rate selection affects fidelity and processing requirements.

Sampling Theorem — A principle stating that a signal must be sampled at least twice its highest frequency component to be accurately reconstructed. This minimum rate is the Nyquist frequency. Violating the theorem causes aliasing. It is fundamental to digital audio.

Scattering Coefficient — A measure of how much sound a surface redirects rather than absorbs or reflects specularly. Higher scattering improves diffusion and reduces coloration. It is important in concert halls, studios, and treatment design. Scattering enhances spatial uniformity.

Schroeder Frequency — The frequency below which room modes dominate and above which sound behaves statistically. It marks the transition between modal and diffuse behavior. The Schroeder frequency guides low-frequency treatment and measurement strategy. Smaller rooms have higher Schroeder frequencies.

Selfextinguishing — A property of materials that stop burning once the ignition source is removed. Self-extinguishing materials improve fire safety in acoustic treatments. They meet building and code requirements. Safety ratings guide material selection.

Self-Noise — The internal noise floor of a microphone or system, expressed in dB-A. It determines how well the device captures subtle details. Lower self-noise improves clarity in quiet environments. It is essential for high-fidelity recording.

Semicircular Canals — Three fluid-filled structures in the inner ear responsible for detecting rotational motion. They contribute to balance and spatial orientation. Although not directly involved in hearing, they influence perception of motion and stability. Their function complements auditory processing.

Send — A routing path that directs part of a signal to an external processor or monitor system. Sends may be pre- or post-fader depending on purpose. They support effects, cue mixes, and parallel processing. Proper send management shapes workflow and clarity.

Sensitivity — A measure of how effectively a microphone or loudspeaker converts input energy into output level. Higher sensitivity yields stronger output for a given input. Sensitivity affects gain staging and noise performance. It is a key specification in device selection.

Sequence Maximum Length — A pseudorandom sequence used in acoustic measurements for system identification. Maximum-length sequences provide broad spectral content and good correlation properties. They support impulse response extraction. MLS techniques are common in room and loudspeaker testing.

Sequence Primitive Root — A mathematical component used to generate pseudorandom sequences such as MLS. Primitive roots ensure maximal sequence length and uniform distribution. They are essential in digital signal processing. Their properties support accurate measurement.

Sequence Quadratic Residue — A sequence based on quadratic residue mathematics used in diffuser design. Quadratic residue sequences determine well depths in QRD diffusers. They produce predictable scattering patterns. Their structure enhances spatial uniformity.

Shaped Sound — Sound that has been filtered or processed to emphasize certain frequencies or characteristics. Shaping improves clarity, reduces noise, or creates artistic effects. It is used in mixing, system tuning, and psychoacoustic design. Shaped sound supports targeted listening experiences.

Sharpness — A psychoacoustic metric describing the high-frequency emphasis of a sound. Sharp sounds feel bright, piercing, or edgy. Sharpness contributes to perceived annoyance. Designers use sharpness metrics in product evaluation.

Shear Wave — A wave in which particles move perpendicular to the direction of propagation. Shear waves occur in solids but not in fluids. They influence vibration, structural acoustics, and seismic behavior. Their speed depends on material rigidity.

Shelving — A type of EQ that boosts or cuts all frequencies above or below a set point. Shelving filters shape tonal balance broadly. They are common in tone controls and mastering. Their gentle curves provide natural-sounding adjustments.

Shield — A conductive barrier that protects audio cables or circuits from electromagnetic interference. Shields reduce hum, RF noise, and crosstalk. Proper grounding enhances shielding effectiveness. Shielding is essential for clean signal transmission.

Shielded Cable — A cable with conductive shielding around its conductors to block interference. Shielded cables improve noise rejection in professional audio. They are essential for long runs and sensitive signals. Proper termination preserves shielding integrity.

Shock Mount — A suspension device that isolates a microphone from mechanical vibration. Shock mounts reduce handling noise, rumble, and stand-borne vibration. They are essential for studio and broadcast work. Proper mounting improves clarity and stability.

Short — An unintended low-resistance connection between two points in a circuit. Shorts cause malfunction, noise, or equipment damage. Troubleshooting identifies and corrects shorts quickly. Proper cable management prevents them.

Shotgun Microphone — A highly directional microphone using an interference tube to reject off-axis sound. Shotgun mics excel at capturing distant or isolated sources. They are widely used in film, broadcast, and field recording. Proper aiming ensures maximum isolation.

Sibilance — Excessive high-frequency energy in speech, especially in "s" and "sh" sounds. Sibilance can be harsh or distracting in recordings. De-essers and microphone choice help control it. Proper technique preserves natural articulation.

Side Lobe — A secondary region of sensitivity or radiation in a polar pattern. Side lobes capture unwanted sound and reduce isolation. They influence feedback risk and tonal consistency. Understanding side lobes guides placement.

Signal — Any electrical or acoustic waveform carrying information. Signals vary in level, frequency, and complexity. Clean signal paths ensure fidelity and clarity. Signal integrity is central to all audio workflows.

Signal Processor — A device or software tool that modifies audio through EQ, dynamics, effects, or filtering. Signal processors shape tone, control levels, and enhance clarity. They are essential in mixing, mastering, and system tuning. Digital processors offer extensive flexibility.

Signal Transmission Level — The strength of a signal as it travels through a system or cable. Proper transmission level ensures compatibility and minimizes noise. Levels must match device specifications. Good gain structure preserves quality.

Signal-to-Noise Ratio — The ratio of desired signal level to background noise level. Higher SNR indicates cleaner, clearer audio. SNR affects intelligibility, fidelity, and dynamic range. Improving SNR is a core goal in system design.

Silencer — A device used to reduce noise in ducts, machinery, or exhaust systems. Silencers use absorptive, reactive, or hybrid designs to target specific frequency ranges. They improve comfort and compliance in HVAC and industrial environments. Proper sizing ensures effective attenuation without excessive airflow resistance.

Simple Harmonic Motion — A periodic motion in which acceleration is proportional to displacement and directed toward equilibrium. It produces sinusoidal waveforms and forms the basis of many acoustic vibrations. Simple harmonic motion describes idealized oscillators such as springs and tuning forks. It underlies much of wave theory.

Simple Harmonic Quantity — Any physical quantity that varies sinusoidally with time. Examples include displacement, velocity, and pressure in harmonic oscillators. These quantities follow predictable mathematical relationships. They are fundamental to acoustic modeling.

Sine Wave — A pure tone with a single frequency and no harmonics. Sine waves form the building blocks of complex sounds through Fourier synthesis. They are used in testing, calibration, and analysis. Their smooth shape represents ideal periodic motion.

SIP (Solo-In-Place) — A solo mode on a mixing console that mutes all other channels while leaving the selected channel routed through the main mix bus. Unlike PFL or AFL, SIP affects the program output, making it useful for evaluating how a channel behaves within the full signal chain, including processing, panning, and effects. Because SIP interrupts the live mix, it is typically used only during soundcheck or in controlled environments. Engineers rely on it to diagnose routing issues, verify processing chains, and assess a channel's true contribution to the mix.

Slap Back — A single, distinct echo occurring shortly after the direct sound. Slap back is common in untreated rooms with reflective surfaces. It adds a vintage or spacious character in music production. Excessive slap back reduces clarity.

Slap Echo — A rapid, discrete echo caused by sound bouncing between parallel surfaces. It produces a sharp, repeating reflection that interferes with intelligibility. Acoustic treatment or diffusion eliminates slap echo. It is a common issue in hard, rectangular rooms.

Sleeve — The ground or shield connection on TS and TRS audio plugs. The sleeve completes the circuit and helps reduce noise. Proper contact ensures clean signal transfer. Sleeves are part of standard phone-plug wiring.

Sliding Fader — A linear control used to adjust audio level smoothly. Sliding faders provide precise, intuitive control in mixers and consoles. Their long travel supports fine adjustments. They are essential in live and studio workflows.

Smoke Developed — A fire-safety rating indicating how much smoke a material produces when burned. Acoustic treatments must meet smoke-development standards for building compliance. Lower values indicate safer materials. This rating guides product selection.

Snake — A bundled multicore cable used to route multiple audio channels between locations. Snakes simplify stage wiring and reduce clutter. They may include analog, digital, or hybrid lines. Proper shielding ensures low noise and reliability.

Soft Room — A room with high absorption and minimal reflections. Soft rooms provide clarity and controlled reverberation. They are ideal for voice recording and critical listening. Excessive softness can make spaces feel acoustically dead.

Solid State — Referring to electronic circuits that use transistors rather than vacuum tubes. Solid-state devices offer reliability, low noise, and compact design. Their sound character is typically clean and stable. They dominate modern audio equipment.

Solo — A function that isolates a channel for monitoring without affecting the main mix. Soloing helps identify issues, adjust tone, and refine performance. It is essential in mixing and troubleshooting. Different consoles offer various solo modes

Solo-In-Place — A solo mode that mutes all other channels while preserving the channel's position in the mix. SIP allows detailed evaluation of processing and effects. It affects the main output, so caution is required during live use. SIP supports precise mix refinement.

Sonar — A system that uses sound waves to detect objects underwater. Sonar measures reflections to determine distance, size, and movement. It is used in navigation, research, and mapping. Acoustic propagation in water enables long-range detection.

Sone — A unit of perceived loudness based on psychoacoustic scaling. One sone corresponds to the loudness of a 1 kHz tone at 40 phons. Sones provide intuitive comparisons of perceived loudness. They support product evaluation and noise studies.

Sonic — Relating to sound waves within the audible range. Sonic characteristics include frequency, amplitude, and timbre. The term is used broadly in acoustics, audio, and perception. Sonic behavior defines how sound is produced and experienced.

Sonic Agglomeration — A clustering of sound sources or reflections that creates a dense acoustic texture. Agglomeration can enhance envelopment or cause muddiness. It depends on room geometry and source distribution. Managing it improves clarity.

Sonic Barrier — A structure designed to block or reduce sound transmission. Sonic barriers are used along highways, industrial sites, and construction zones. Their effectiveness depends on height, mass, and placement. They improve environmental noise control.

Sonic Boom — A shock wave produced when an object travels faster than the speed of sound. Sonic booms create intense, impulsive noise and pressure changes. They are characteristic of supersonic aircraft. Regulations limit boom exposure in populated areas.

Sonic Delay Line — A device or medium that delays sound using acoustic propagation rather than electronics. Examples include tubes, springs, or mechanical paths. Sonic delay lines were used in early effects units. They create distinctive tonal coloration.

Sonic Drilling — A drilling method that uses high-frequency vibration to penetrate soil or rock. Sonic energy reduces friction and improves drilling efficiency. It is used in geotechnical and environmental applications. Acoustic resonance enhances performance.

Sonic Frequency — Any frequency within the audible range of human hearing, typically 20 Hz to 20 kHz. Sonic frequencies define pitch, timbre, and tonal balance. They are central to audio design and perception. Measurement tools analyze sonic content.

Sonic Soldering — A soldering technique that uses ultrasonic vibration to bond materials. Sonic energy improves wetting and reduces oxidation. It is used for delicate or specialized components. The process enhances reliability and precision.

Sonic Speed — The speed at which sound travels through a medium. Sonic speed depends on temperature, density, and elasticity. In air at room temperature, it is approximately 343 m/s. It influences propagation, refraction, and acoustic timing.

Sonic Wave — A mechanical wave that propagates through a medium within the audible frequency range. Sonic waves involve alternating compression and rarefaction. They carry energy and information. Their behavior defines acoustic phenomena.

Sonics — The study and application of sound behavior, production, and perception. Sonics encompasses acoustics, psychoacoustics, and audio engineering. It examines how sound interacts with environments and systems. The field supports design, analysis, and innovation.

Sonoluminescence — The emission of light from collapsing bubbles in a liquid when excited by intense sound. This phenomenon involves extreme pressure and temperature. It is studied in physics, ultrasonics, and fluid dynamics. Sonoluminescence illustrates nonlinear acoustic effects.

Sound — A mechanical vibration that propagates through a medium as a pressure wave. Sound conveys information, emotion, and environmental cues. Its characteristics include frequency, amplitude, and timbre. Sound is central to communication and perception.

Sound Absorption — The process by which materials convert sound energy into heat, reducing reflections. Absorption improves clarity, reduces reverberation, and controls noise. Materials vary in effectiveness across frequencies. Proper absorption design shapes room acoustics.

Sound Absorption Coefficient — A measure of how effectively a material absorbs sound at a given frequency. Values range from 0 (no absorption) to 1 (total absorption). Coefficients guide material selection and acoustic modeling. They are essential in room design.

Sound Attenuation — The reduction of sound level as it travels through a medium or barrier. Attenuation results from absorption, scattering, and geometric spreading. It is central to noise control and system design. Proper attenuation improves comfort and privacy.

Sound Chain — The complete path a signal follows from source to listener. The chain includes microphones, processors, amplifiers, and loudspeakers. Each stage affects quality and clarity. Managing the sound chain ensures reliable performance.

Sound Energy — The total acoustic energy carried by a sound wave. It depends on pressure, particle velocity, and duration. Sound energy influences loudness and impact. Measuring it supports analysis and design.

Sound Energy Density — The amount of sound energy per unit volume in a medium. Higher density corresponds to stronger acoustic fields. It is used in modeling, measurement, and environmental assessment. Energy density helps describe sound intensity and distribution.

Sound Energy Flux — The rate at which sound energy passes through a unit area. It describes the directional flow of acoustic energy in a sound field. Flux helps identify how energy moves from sources to receivers. It is central to intensity-based measurements.

Sound Energy Flux Density — The sound energy flux per unit area, representing how much acoustic energy flows through a surface. It combines pressure and particle velocity information. Flux density supports detailed analysis of sound fields. It is used in intensity probes and environmental acoustics.

Sound Exposure — The total sound energy a listener receives over a period of time. Exposure accounts for both level and duration. It is used in hearing-risk assessment and regulatory compliance. High exposure increases the risk of hearing damage.

Sound Exposure Dose — A measure of cumulative sound exposure expressed as a percentage of a maximum allowable dose. It reflects how much of a safe daily limit has been used. Dose metrics guide occupational hearing protection. They help prevent long-term auditory damage.

Sound Exposure Level — A single-number metric that expresses the total energy of a sound event normalized to one second. SEL allows comparison of events with different durations. It is widely used in environmental noise assessment. Higher SEL indicates greater energy impact.

Sound Field — The distribution of sound pressure and particle velocity in a space. Sound fields may be direct, reverberant, diffuse, or near-field. Their characteristics influence clarity, loudness, and tonal balance. Understanding sound fields guides acoustic design.

Sound Insulation — The ability of a structure to block sound transmission between spaces. Good insulation improves privacy and reduces noise intrusion. It depends on mass, stiffness, and construction quality. Proper insulation is essential in buildings and studios.

Sound Intensity — The acoustic power passing through a unit area in a specific direction. Intensity combines pressure and particle velocity, making it a vector quantity. It reveals energy flow and source direction. Intensity measurements support diagnostics and noise control.

Sound Intensity Level — A logarithmic measure of sound intensity relative to a reference value. It expresses intensity in decibels. Higher levels indicate stronger energy flow. It complements sound pressure level in analysis.

Sound Intensity Level Difference — The difference in intensity level between two points or directions. It helps identify source direction, reflections, and energy distribution. This metric is used in binaural hearing and system tuning. Differences reveal spatial characteristics.

Sound Intensity Probe — A measurement device using two closely spaced microphones to capture both pressure and particle velocity. Intensity probes determine sound intensity, direction, and energy flow. They are essential in noise diagnostics and acoustic research. Proper calibration ensures accuracy.

Sound Intensity Vector — A vector representing the magnitude and direction of sound intensity. It shows how acoustic energy moves through space. Vector analysis helps identify sources and reflections. It is used in advanced measurement and modeling.

Sound Isolation — The reduction of sound transmission between spaces through construction and design. Isolation relies on mass, decoupling, and airtightness. Good isolation improves privacy and recording quality. It is a core principle of architectural acoustics.

Sound Level — The magnitude of sound pressure expressed in decibels. Sound level indicates loudness and exposure risk. It varies with distance, environment, and source characteristics. Accurate measurement supports regulation and design.

Sound Level Meter — An instrument used to measure sound pressure level with standardized weighting and time responses. SLMs are essential for environmental, occupational, and product testing. They provide objective, repeatable measurements. Compliance requires calibration and proper technique.

Sound Power — The total acoustic energy emitted by a source per unit time. Sound power is independent of distance or environment. It is used to compare noise emissions across products. Manufacturers report sound power for regulatory compliance.

Sound Power Directivity — The variation of sound power radiated in different directions. Directivity patterns reveal how sources distribute energy. They influence coverage, reflections, and system design. Understanding directivity improves prediction and control.

Sound Power Level — A logarithmic measure of sound power relative to a reference value. It expresses a source's inherent noise output. Sound power level is used in product labeling and environmental modeling. It differs from sound pressure level, which depends on location.

Sound Pressure — The local deviation from ambient atmospheric pressure caused by a sound wave. Sound pressure determines loudness and waveform shape. It is the primary quantity measured in acoustics. Pressure varies with distance and environment.

Sound Pressure Amplitude — The maximum deviation of sound pressure from its equilibrium value. Amplitude determines perceived loudness and energy. Higher amplitude produces stronger acoustic effects. It is central to waveform analysis.

Sound Pressure Level — A logarithmic measure of sound pressure relative to a reference value, expressed in decibels. SPL is the most common metric for describing loudness. It supports regulation, measurement, and system tuning. Weighting filters adjust SPL to reflect human hearing.

Sound Probe — A measurement device used to capture sound pressure or intensity in specific locations. Probes support detailed spatial analysis. They are used in research, diagnostics, and product testing. Proper placement ensures accurate results.

Sound Quality Metrics — Quantitative measures describing the perceptual attributes of sound, such as loudness, sharpness, roughness, and fluctuation strength. These metrics reflect human response rather than purely physical properties. They guide product design and evaluation. Sound quality metrics support psychoacoustic optimization.

Sound Receiver — Any device or location that receives sound energy, such as a microphone or listener position. Receiver characteristics influence measurement and perception. Proper receiver placement ensures accurate capture. Receivers are central to system evaluation.

Sound Reduction Index — A measure of how much sound a building element reduces transmission across it. Higher SRI values indicate better isolation. It is frequency-dependent and used in laboratory testing. SRI supports material comparison and design.

Sound Reinforcement — The use of audio systems to amplify and distribute sound for an audience. Reinforcement systems include microphones, processors, amplifiers, and loudspeakers. Their design ensures clarity, coverage, and appropriate loudness. Good reinforcement enhances natural sound.

Sound Spectrograph — A device or software tool that displays sound frequency content over time. Spectrographs reveal harmonics, transients, and noise patterns. They support analysis, diagnostics, and research. Spectrograms are widely used in acoustics and audio engineering.

Sound Strength — A measure of how much a room amplifies or attenuates sound relative to a reference distance. Sound strength reflects room gain and energy distribution. It is used in concert hall evaluation. Higher values indicate stronger perceived loudness.

Sound Transmission Class — A single-number rating describing a partition's ability to block airborne sound. STC values are based on standardized laboratory tests. Higher STC indicates better isolation. STC is widely used in building codes and design.

Sound Transmission Coefficient — The fraction of incident sound energy that passes through a material. Lower coefficients indicate better isolation. It varies with frequency and construction. It is used in modeling and material evaluation.

Sound Transmission Loss — The reduction in sound level as it passes through a building element. TL is frequency-dependent and measured in decibels. Higher TL indicates better isolation. It is fundamental to architectural acoustics.

Sound Wave — A mechanical disturbance that propagates through a medium as alternating compressions and rarefactions. Sound waves carry energy and information. Their behavior depends on frequency, amplitude, and environment. They form the basis of all acoustics.

Sound Waves — The plural form of sound wave, referring to multiple propagating disturbances. Sound waves interact through reflection, refraction, diffraction, and interference. Their combined behavior shapes acoustic environments. Understanding them supports design and analysis.

Sound-Energy Flux — Another term for sound energy flux, describing the rate of acoustic energy flow through a surface. It emphasizes directional energy transport. Flux analysis supports intensity-based diagnostics. It is used in environmental and industrial acoustics.

Sounding — The process of measuring depth or probing an environment using sound. Soundings are common in marine navigation and geological surveys. They rely on echo timing and propagation speed. Accurate sounding supports mapping and safety.

Sounding Rocket — A rocket used to carry scientific instruments into suborbital space for short-duration measurements. Sounding rockets study atmospheric acoustics, ionospheric behavior, and space physics. They provide data unavailable from ground-based systems. Their brief flights support targeted research.

Soundstage — The perceived spatial arrangement of sound sources in a recording or playback system. A strong soundstage provides clear width, depth, and localization cues. It depends on microphone technique, room acoustics, and loudspeaker placement. Good soundstage enhances realism and listener immersion.

Source Level — The sound power or pressure level emitted directly from a source before propagation effects. It represents the inherent loudness of the source. Source level is used in environmental modeling and underwater acoustics. Accurate measurement supports prediction of received levels.

Source Room — The room in which a sound source is located during transmission testing. Its acoustics influence measured sound levels and transmission behavior. Standardized conditions ensure reliable comparison. Source-room characteristics affect airborne sound isolation results.

Spatial Decay Rate — The rate at which sound level decreases with distance in a room. It reflects absorption, scattering, and room geometry. Higher decay rates indicate more absorptive spaces. Spatial decay is central to clarity and speech intelligibility.

Spatial Impulse Response — An impulse response measured at multiple spatial positions to capture directional and spatial characteristics. It reveals reflections, diffusion, and room geometry effects. Spatial IRs support 3D audio rendering and acoustic modeling. They provide richer information than single-point measurements.

Spatial Room Impulse Response — A room impulse response recorded with spatially sensitive microphones or arrays. It captures both temporal and directional information. SRIRs support binaural rendering, auralization, and immersive audio. They are essential in advanced room-acoustic analysis.

Speaker Level — A high-power audio signal intended to drive loudspeakers directly. Speaker-level signals require robust cables and connectors. They are much stronger than line- or mic-level signals. Proper handling prevents overload and equipment damage.

Speakon — A locking connector designed for high-power speaker connections. Speakon connectors provide secure, reliable contact and prevent accidental disconnection. They are standard in professional sound reinforcement. Their design supports safe, high-current operation.

Specific Airflow Resistance — A measure of how much a porous material resists airflow. It strongly influences absorption performance, especially at mid- and high-frequencies. Proper airflow resistance ensures effective acoustic treatment. It is a key parameter in material design.

Spectral Balance — The relative distribution of energy across the frequency spectrum. Good spectral balance ensures natural, clear sound. Imbalances cause harshness, muddiness, or thinness. Engineers adjust spectral balance through EQ and mixing.

Spectral Splatter — Unwanted spreading of energy into adjacent frequencies, often caused by clipping or rapid transitions. Splatter reduces clarity and introduces distortion. It is common in RF systems and digital audio. Proper filtering and gain staging minimize it.

Spectrogram — A visual representation of sound showing frequency over time with amplitude indicated by color or intensity. Spectrograms reveal harmonics, transients, and noise patterns.

They support analysis, diagnostics, and research. They are widely used in acoustics and audio engineering.

Spectrum — The distribution of a signal's energy across frequencies. Spectra reveal tonal content, noise characteristics, and system behavior. They are fundamental to analysis and design. Spectrum shape defines timbre and clarity.

Spectrum Analyzer — A device or software tool that displays the frequency spectrum of a signal. Spectrum analyzers help identify peaks, noise, and distortion. They are essential in system tuning and diagnostics. Modern analyzers offer real-time, high-resolution displays.

Spectrum Density — A measure of how signal power is distributed per unit frequency. It supports detailed spectral analysis and comparison. Density plots reveal noise floors and tonal components. They are used in research and diagnostics.

Specular Reflection — A mirror-like reflection where sound bounces off a surface at an angle equal to its incidence. Specular reflections preserve waveform shape and direction. They influence clarity, imaging, and coloration. Controlling them is key in room design.

Specular Reflections — Multiple mirror-like reflections from smooth surfaces. They create strong early reflections that affect intelligibility and imaging. Excessive specular reflections cause flutter echo and coloration. Treatment or diffusion reduces their impact.

Specularity — The degree to which a surface reflects sound in a mirror-like manner. High specularity produces strong, directional reflections. Low specularity increases diffusion. Specularity influences room acoustics and treatment strategy.

Speech Intelligibility — The clarity with which speech can be understood in a given environment. Intelligibility depends on reverberation, noise, frequency balance, and direct-to-reverberant ratio. High intelligibility is essential for classrooms, auditoriums, and communication systems. Metrics such as STI quantify it.

Speech Interference Level — A metric that averages noise levels in key speech-frequency bands. SIL predicts how background noise affects speech communication. Higher SIL indicates greater masking. It is used in environmental and building acoustics.

Speech Privacy Class — A rating that measures how well a space prevents overheard speech from being understood. SPC considers background noise, masking, and partition performance. Higher values indicate better privacy. It is used in offices, healthcare, and secure environments.

Speech Transmission Index — A standardized measure of speech intelligibility based on modulation transfer. STI ranges from 0 (poor) to 1 (excellent). It accounts for noise, reverberation, and distortion. STI is widely used in room design and system evaluation.

Speed of Sound — The rate at which sound waves travel through a medium. Speed depends on temperature, density, and elasticity. In air at room temperature, it is about 343 m/s. It influences timing, wavelength, and propagation behavior.

Spherical Divergence — The spreading of sound energy as it radiates outward from a point source. Spherical divergence causes level to drop by 6 dB per doubling of distance. It is fundamental to propagation modeling. Deviations occur in directional or reverberant fields.

Spherical Harmonics — Mathematical functions used to describe angular variations in wavefields. They support 3D audio, beamforming, and acoustic modeling. Spherical harmonics represent complex spatial patterns efficiently. They are central to Ambisonics and spatial acoustics.

Spherical Wave — A sound wave that radiates outward uniformly in all directions from a point source. Spherical waves exhibit geometric spreading and predictable decay. They approximate real-world sources at moderate distances. Their behavior underlies many acoustic models.

Spill — Unwanted sound from nearby sources entering a microphone. Spill reduces isolation and complicates mixing. It is influenced by microphone pattern, placement, and room acoustics. Managing spill improves clarity and control.

SPL — Sound Pressure Level, a logarithmic measure of sound pressure relative to a reference. SPL indicates loudness and exposure risk. It is the most common acoustic metric. Weighting filters adjust SPL to reflect human hearing.

Splaying — The intentional angling of walls or surfaces to reduce parallel reflections. Splaying prevents flutter echo and improves diffusion. It is common in studios and performance spaces. Proper geometry enhances clarity and spatial uniformity.

Splitter — A device that divides an audio signal into multiple outputs. Splitters support recording, monitoring, and routing. Passive and active designs offer different performance characteristics. Proper splitting preserves signal integrity.

Spring Reverb — A reverb effect created by sending audio through vibrating metal springs. Spring reverbs produce a distinctive metallic, boing-like character. They are common in guitar amplifiers and vintage gear. Their coloration is used creatively in music production.

Squawker — A loudspeaker driver designed to reproduce midrange frequencies. Squawkers bridge the gap between woofers and tweeters. Their performance influences clarity and vocal presence. They are common in multi-way loudspeaker systems.

Squelch — A circuit that mutes audio output when signal strength falls below a threshold. Squelch prevents noise and static in wireless systems. Proper adjustment ensures clean operation. It is essential in RF communication.

Stage Rumble — Low-frequency vibration transmitted through a stage into microphones and stands. Stage rumble reduces clarity and can cause feedback. Isolation mounts and filtering help control it. It is common in live sound environments.

Standing Wave — A stationary pattern formed when incident and reflected waves combine. Standing waves create peaks and nulls in frequency response. They are most problematic at low frequencies. Managing them improves tonal accuracy.

Standing Wave Ratio — A measure of how strongly standing waves are present in a system, often used in transmission lines. High SWR indicates poor impedance matching and strong reflections. Low SWR ensures efficient energy transfer. It is important in RF and acoustic systems.

Static Pressure — The uniform, non-moving component of pressure in a fluid such as air. In acoustics, static pressure forms the baseline around which sound-induced pressure fluctuations occur. HVAC systems rely on static pressure to move air effectively. Proper control prevents noise, turbulence, and performance loss.

Steady Noise — Noise that remains relatively constant in level and spectral content over time. Examples include fan hum, HVAC noise, and continuous machinery operation. Steady noise is easier to measure and predict than fluctuating noise. It forms the basis for many environmental and occupational assessments.

Steady-State — A condition in which system variables remain constant over time despite ongoing processes. In acoustics, steady-state sound fields occur when reflections and energy distribution stabilize. Measurements often require steady-state conditions for accuracy. It contrasts with transient or time-varying behavior.

Stereo — A two-channel audio format that creates spatial imaging through differences in level, timing, and frequency content between left and right channels. Stereo enhances realism and immersion. Proper playback requires accurate speaker placement. It is the dominant format for music and media.

Stereo Image — The perceived spatial placement of sounds within a stereo field. A strong stereo image provides clear localization, width, and depth. It depends on microphone technique, mix decisions, and playback geometry. Good imaging enhances realism and listener engagement.

Stereo Microphone — A microphone or microphone pair designed to capture stereo sound. Techniques include XY, ORTF, MS, and spaced pairs. Stereo microphones preserve spatial cues and ambience. They are widely used in field recording, music production, and broadcast.

Stereo Return — A mixer input designed to receive stereo signals from effects processors or playback devices. Stereo returns simplify routing and level control. They maintain left-right imaging without requiring two separate channels. They are common in live and studio consoles.

Stiffness — A material's resistance to deformation under force. In acoustics, stiffness influences resonance, vibration, and transducer behavior. High stiffness raises resonant frequencies and improves structural stability. It is a key parameter in loudspeaker and enclosure design.

Stochastic Noise — Noise characterized by random, probabilistic behavior rather than deterministic patterns. Stochastic noise includes broadband environmental noise and many natural sound sources. It is analyzed statistically rather than through discrete components. Such noise is common in environmental and industrial acoustics.

Strouhal Number — A dimensionless number describing oscillating flow phenomena such as vortex shedding. In acoustics, it helps predict flow-induced noise in pipes, fans, and aerodynamic structures. Strouhal scaling supports aeroacoustic modeling. It links frequency, velocity, and characteristic length.

Structural Damping — The dissipation of vibrational energy within a structure due to internal friction or material properties. Higher damping reduces resonance and vibration amplitude. It improves comfort, stability, and acoustic isolation. Damping treatments enhance building and product performance.

Structure-Borne Sound — Sound transmitted through solid materials rather than through air. Examples include footsteps, machinery vibration, and plumbing noise. Structure-borne sound often bypasses airborne barriers. Isolation, decoupling, and damping help control it.

Stud — A vertical framing member in building construction. Studs transmit sound structurally and influence wall isolation. Double-stud or staggered-stud designs improve acoustic performance. Stud spacing affects resonance and stiffness.

Sub — A shorthand term for subwoofer, referring to a loudspeaker dedicated to low-frequency reproduction. Subs extend system bandwidth and provide impact. Proper placement and crossover settings ensure smooth integration. They are essential in cinema, music, and reinforcement systems.

Sub-Bottom Profiling — An acoustic technique used to image sediment layers beneath the seafloor. It uses low-frequency sound pulses to penetrate geological structures. Sub-bottom profiling supports marine research, archaeology, and engineering. It relies on reflection timing and propagation speed.

Subgroup — A mixer routing feature that combines multiple channels into a single controllable group. Subgroups simplify level control, processing, and automation. They are useful for drums, vocals, or instrument sections. Subgrouping improves workflow and mix organization.

Subharmonic — A frequency component that is an integer fraction of a fundamental frequency. Subharmonics add depth and richness to sound. They occur naturally in nonlinear systems or through processing. Subharmonic synthesis enhances bass content.

Submaster — A secondary master control that adjusts the level of a subgroup or section of a mix. Submasters allow coordinated control of related channels. They support efficient mixing and scene changes. Proper use maintains balance and dynamics.

Submixer — A smaller mixer used to combine multiple sources before feeding a main console. Submixers manage complex setups such as drum kits or multi-mic instruments. They reduce channel count on the main board. Submixing improves organization and flexibility.

Subs — A plural shorthand for subwoofers. Multiple subs improve low-frequency coverage and reduce modal issues. Their placement strongly affects room response. Proper alignment ensures smooth, powerful bass.

Subsonic — Referring to frequencies below the audible range or speeds below the speed of sound. Subsonic content may cause rumble or mechanical noise. Filters remove unwanted subsonic energy. Subsonic speeds avoid shock waves and sonic booms.

Subsonic Flow — Fluid flow at speeds below the speed of sound. Subsonic flow exhibits smooth, predictable aerodynamic behavior. It contrasts with supersonic flow, which produces shock waves. Subsonic modeling is common in HVAC and environmental acoustics.

Subwoofer — A loudspeaker designed to reproduce low-frequency sound, typically below 120 Hz. Subwoofers add depth, impact, and realism to audio systems. Their performance depends on enclosure design, placement, and crossover integration. They are essential in cinema, music, and reinforcement.

Summed Mono — A signal created by combining left and right stereo channels into a single mono output. Summed mono ensures compatibility for single-speaker playback. It reveals phase issues and imbalances. Many systems use summed mono for distributed audio.

Supercardioid Microphone — A directional microphone with a narrow front pickup and a small rear lobe. Supercardioids offer strong isolation and feedback resistance. They are ideal for stage, film, and broadcast. Placement must account for the rear pickup zone.

Superposition — The principle that multiple waves can combine by simple addition of their amplitudes. Superposition explains interference, reinforcement, and cancellation. It applies to linear acoustic systems. Complex sound fields result from overlapping waves.

Supersonic — Referring to speeds greater than the speed of sound. Supersonic motion produces shock waves and intense aerodynamic noise. It is central to aerospace acoustics. Supersonic behavior differs fundamentally from subsonic flow.

Supersonic Compressor — A device or aerodynamic structure that compresses air at supersonic speeds. Supersonic compressors appear in advanced propulsion and research applications. They generate strong shock waves and high noise levels. Their design requires specialized aeroacoustic modeling.

Supersonic Diffuser — A diffuser designed to slow supersonic flow to subsonic speeds. It manages shock waves and pressure recovery. Supersonic diffusers are used in wind tunnels and propulsion systems. Their geometry controls flow stability.

Supersonic Flow — Fluid flow at speeds exceeding the speed of sound. Supersonic flow produces shock waves, rapid pressure changes, and intense noise. It is central to high-speed aerodynamics and aeroacoustics. Modeling requires nonlinear equations.

Supersonic Nozzle — A nozzle designed to accelerate flow to supersonic speeds. These nozzles shape shock patterns and jet noise. They are used in rockets, turbines, and research. Their acoustics are complex and high-energy.

Supersonics — The study of phenomena associated with supersonic speeds. It includes shock waves, aerodynamic heating, and high-intensity noise. Supersonics intersects with aeroacoustics and fluid dynamics. It supports aerospace design and research.

Surround Sound — A multichannel audio format that places sound around the listener for immersive experience. Surround systems use rear, side, and sometimes height channels. They enhance spatial realism in cinema, gaming, and home theater. Proper calibration ensures seamless envelopment.

Sweepable EQ — An equalizer with adjustable center frequency for each band. Sweepable EQs allow precise tonal shaping and problem correction. They are common in mixing consoles and outboard gear. Their flexibility surpasses fixed-frequency EQs.

Synthesizer — An electronic instrument that generates sound through oscillators, filters, and modulation. Synthesizers create a vast range of tones, from natural to abstract. They are central to modern music production. Their flexibility supports creative sound design.

T60 — A measure of reverberation time indicating how long it takes for sound to decay by 60 dB. T60 defines room acoustics and influences clarity, warmth, and envelopment. Ideal values vary by room function. It is a core metric in architectural acoustics.

Talkback — A communication system that allows engineers or directors to speak to performers or crew through headphones or monitors. Talkback enables coordination during recording, broadcast, and live production. It typically uses a dedicated microphone and routing path. Proper talkback setup improves workflow and reduces interruptions.

Tangential Mode — A room mode involving reflections from two pairs of parallel surfaces. Tangential modes are less dominant than axial modes but still influence low-frequency response. Their frequencies depend on room dimensions. They contribute to modal density and overall bass smoothness.

Tape Return — A mixer input designed to receive playback from a recording device. Tape returns allow monitoring of recorded tracks without repatching. They support overdubbing, editing, and mix review. Modern systems use digital equivalents but retain the same workflow concept.

Target Strength — A measure of how strongly an underwater object reflects sonar signals. Higher target strength indicates a more detectable object. It depends on size, shape, material, and orientation. Target strength is central to sonar performance and classification.

TDS — Time-Delay Spectrometry, a measurement technique that separates direct sound from reflections using time-windowed analysis. TDS provides accurate frequency response in reverberant spaces. It is useful for loudspeaker testing and room evaluation. The method improves clarity in complex environments.

TEF — Time-Energy-Frequency analysis, a measurement system that visualizes acoustic behavior across time, energy, and frequency domains. TEF supports detailed room and loudspeaker evaluation. It reveals reflections, decay patterns, and spectral balance. TEF is widely used in professional acoustics.

Temporal Masking — A psychoacoustic effect where one sound makes another sound harder to hear when they occur close in time. Forward masking occurs when a loud sound masks a following sound; backward masking occurs when it masks a preceding one. Temporal masking influences perception, coding, and intelligibility. It is central to audio compression algorithms.

Temporary Threshold Shift — A short-term elevation in hearing threshold after noise exposure. Hearing typically recovers over hours or days. Repeated temporary shifts can lead to permanent damage. TTS is an early warning sign of excessive exposure.

THD — Total Harmonic Distortion, a measure of unwanted harmonic content added by a device. Lower THD indicates cleaner, more accurate reproduction. THD is used to evaluate amplifiers, converters, and transducers. It is expressed as a percentage or decibel value.

Thermoacoustics — The study of how heat and sound interact in fluids and solids. Thermoacoustic effects include sound generation from temperature gradients and heat transport by acoustic waves. Applications include engines, refrigerators, and advanced research. The field blends acoustics and thermodynamics.

Thiele–Small Parameters — A set of electromechanical parameters describing loudspeaker driver behavior. These parameters predict resonance, compliance, efficiency, and enclosure performance. They guide cabinet design and system tuning. T-S parameters are fundamental to loudspeaker engineering.

Threshold of Audibility — The minimum sound level a typical human can detect at a given frequency. It varies with frequency and listening conditions. Threshold curves define equal-loudness contours. They form the basis of psychoacoustic weighting.

Threshold of Detectability — The lowest level at which a sound can be reliably detected in the presence of noise. It depends on masking, frequency, and listener sensitivity. Detectability thresholds guide communication system design. They differ from pure audibility thresholds.

Threshold of Discomfort — The sound level at which listeners begin to feel physical discomfort. It typically occurs around 100–120 dB SPL. This threshold varies with frequency and individual sensitivity. It guides safe exposure limits.

Threshold of Feeling — The level at which sound produces noticeable physical sensations such as vibration or pressure. It occurs at high SPLs, often above 120 dB. Low frequencies reach this threshold sooner. It is relevant in high-intensity acoustics.

Threshold of Hearing — The quietest sound a typical listener can perceive, usually defined as 0 dB SPL at 1 kHz. Thresholds vary with frequency and individual hearing. They form the basis of loudness perception. Hearing tests measure deviations from this threshold.

Threshold of Pain — The sound level at which sound becomes physically painful, typically around 130–140 dB SPL. Exposure at this level risks immediate damage. It is used in safety standards and hazard assessment. The threshold varies among individuals.

Threshold Sensitivity — The minimum input level required for a device or system to respond. In microphones, it refers to the quietest sound they can detect. In electronics, it defines activation or triggering levels. High sensitivity improves detection of subtle signals.

THX — A certification and performance standard for audio and cinema systems. THX ensures consistent playback quality, clarity, and spatial accuracy. It covers equipment, room design, and calibration. THX branding indicates adherence to strict specifications.

Timbral — Relating to timbre, the tonal quality that distinguishes different sound sources. Timbral characteristics include harmonic content, attack, and spectral shape. They define the identity of instruments and voices. Timbral balance is central to mixing and perception.

Timbre — The tonal character of a sound that differentiates one source from another, even at the same pitch and loudness. Timbre arises from harmonics, transients, and spectral distribution. It shapes musical identity and emotional impact. Accurate reproduction preserves timbre.

Time-Delay Spectrometry — A measurement technique that isolates direct sound from reflections using time-windowed analysis. It provides accurate frequency response in reverberant spaces. TDS is valuable for loudspeaker testing and room tuning. It improves clarity in complex environments.

Time-of-Flight — The time it takes for a sound wave to travel from source to receiver. Time-of-flight measurements determine distance, location, and propagation speed. They are used in sonar, acoustics, and sensing. Accurate timing supports spatial analysis.

Time-Stretching — A digital process that changes the duration of audio without altering pitch. Time-stretching supports editing, synchronization, and creative effects. Advanced algorithms preserve timbre and clarity. It is widely used in music production and post-processing.

Timoshenko Beam Theory — A structural model that accounts for shear deformation and rotational inertia in beams. It provides more accurate predictions than classical beam theory at high frequencies or short lengths. Timoshenko theory is used in vibration and structural acoustics. It supports advanced modeling of panels and enclosures.

Tip — The end contact of a TS or TRS plug that carries the primary signal. Tip connections vary by wiring standard. Proper contact ensures clean signal transfer. The tip is part of standard phone-plug nomenclature.

Tonal — Describing sound with clear pitch or harmonic structure. Tonal components stand out against noise and are easily perceived. Tonal noise is often more annoying than broadband noise. Tonality influences perception and regulation.

Tonal Audibility — A measure of how easily a tonal component can be heard within noise. It depends on masking, frequency, and level. High tonal audibility increases perceived annoyance. It is used in product evaluation and environmental acoustics.

Tone — A sound with a specific pitch and harmonic structure. Tones may be pure or complex. They form the basis of music and auditory perception. Tone quality influences timbre and emotional response.

Tone Burst — A short, controlled burst of tonal energy used in testing. Tone bursts reveal transient behavior, resonance, and distortion. They support loudspeaker and room analysis. Their duration and frequency are precisely defined.

Tone Control — A simple EQ circuit that adjusts bass and treble balance. Tone controls shape overall spectral character. They are common in consumer audio and musical instruments. Their broad adjustments provide intuitive shaping.

Tone Squelch — A system that uses a subaudible tone to control muting in communication devices. Receivers unmute only when the correct tone is detected. Tone squelch reduces interference and cross-talk. It is widely used in radio systems.

Total Harmonic Distortion — A measure of harmonic distortion produced by a device. THD quantifies how much unwanted harmonic energy is added to a signal. Lower THD indicates cleaner reproduction. It is a key specification for amplifiers and converters.

Transducer — A device that converts one form of energy into another, such as sound into electricity or vice versa. Microphones, loudspeakers, and sensors are all transducers. Their design influences sensitivity, bandwidth, and distortion. Transducers are fundamental to audio systems.

Transducer Gain — The ratio of output energy to input energy in a transducer. Gain reflects efficiency and sensitivity. Higher gain improves output level but may increase noise. It is important in system design and matching.

Transfer Function — A mathematical description of how a system modifies input signals across frequency. Transfer functions reveal gain, phase shift, and resonance. They are essential in system tuning, modeling, and diagnostics. Accurate measurement supports optimization.

Transfer Matrix Method — A mathematical technique used to model how sound propagates through layered media or duct systems. It represents each layer with a matrix describing its acoustic behavior. By multiplying matrices, complex systems can be analyzed efficiently. The method is widely used in muffler design, building acoustics, and transducer modeling.

Transformer — An electrical device that transfers energy between circuits through magnetic induction. Transformers provide isolation, impedance matching, and voltage conversion. In audio, they reduce noise and improve signal integrity. High-quality transformers preserve frequency response and minimize distortion.

Transient — A short, non-steady burst of sound energy, such as a drum hit or consonant. Transients define attack, clarity, and articulation. They reveal system speed and accuracy. Capturing transients cleanly is essential for natural-sounding audio.

Transient Response — A system's ability to reproduce rapid changes in sound. Good transient response preserves detail, attack, and clarity. Poor response smears or dulls transients. It is a key indicator of loudspeaker and microphone performance.

Transients — Brief, high-energy events that occur at the start of sounds. They contain important timing and timbral cues. Transients challenge audio systems due to their speed and amplitude. Accurate reproduction enhances realism.

Transmission Loss — The reduction in sound level as it passes through a building element. TL is frequency-dependent and measured in decibels. Higher TL indicates better isolation. It is fundamental to architectural acoustics.

Transmit Current Response — A measure of how a transducer's output varies with the electrical current driving it. It reveals efficiency, resonance, and distortion. Current-based response is useful in underwater acoustics and specialized transducers. It complements voltage-based measurements.

Transmit Voltage Response — A measure of a transducer's output relative to the applied voltage. It describes sensitivity, bandwidth, and resonance. Voltage response is a standard metric in loudspeaker and hydrophone testing. It supports system design and calibration.

Transmitter — A device that sends audio or data wirelessly using radio frequency signals. Transmitters are used in wireless microphones, communication systems, and monitoring. Their performance depends on power, frequency coordination, and antenna design. Reliable transmission ensures clean, dropout-free audio.

Transonic — Referring to speeds near the speed of sound, typically Mach 0.8–1.2. Transonic flow exhibits mixed subsonic and supersonic regions. It produces strong aerodynamic noise and shock formation. Transonic behavior is important in aeroacoustics and aircraft design.

Transonic Flow — Fluid flow in which local velocities transition between subsonic and supersonic. This regime produces complex shock patterns and high noise levels. It challenges modeling and stability. Transonic flow is central to high-speed aerodynamics.

Transonic Speed — The speed range around the speed of sound where transonic effects occur. Objects at transonic speed experience rapid changes in drag and noise. It marks the onset of shock waves. Engineers design structures to manage these effects.

Transparency — The ability of an audio system to reproduce sound without coloration or distortion. Transparent systems preserve detail, timbre, and spatial cues. Transparency is prized in mastering, monitoring, and high-fidelity playback. It reflects accuracy rather than tonal character.

Transponder — A device that receives a signal and automatically transmits a response. Transponders are used in navigation, communication, and identification systems. In acoustics, they support underwater tracking and measurement. Their behavior depends on frequency and coding.

Transverse Wave — A wave in which particles move perpendicular to the direction of propagation. Transverse waves occur in solids and on surfaces but not in fluids. They influence vibration and structural acoustics. Their speed depends on material rigidity.

Traveling Frequency — The frequency component of a wave as it propagates through a medium. Traveling frequencies determine pitch and spectral content. They interact with boundaries and materials. Understanding them supports analysis and design.

Treble — The high-frequency portion of the audio spectrum, typically above 4 kHz. Treble contributes brightness, clarity, and detail. Excessive treble sounds harsh; insufficient treble sounds dull. Treble control shapes tonal balance.

Tri-Amp — A loudspeaker system using three separate amplifiers for low, mid, and high-frequency drivers. Tri-amping improves control, efficiency, and clarity. It requires active crossovers and precise tuning. Professional systems often use tri-amp configurations.

Triboelectric Noise — Noise generated when friction between materials creates electrical charge variations. In audio cables, it appears as crackling or rustling when cables move. Proper shielding and cable design reduce triboelectric noise. It is a concern in sensitive microphone lines.

Trim Control — A gain adjustment used to set input level before further processing. Trim ensures proper headroom and noise performance. It is essential in mixing consoles and interfaces. Correct trim prevents clipping and distortion.

Tube Traps — Cylindrical acoustic absorbers designed to control low-frequency resonance and reflections. Tube traps combine absorption and diffusion. They are effective in studios and listening rooms. Placement influences bass smoothness and clarity.

Tuning Frequency — The resonant frequency of a loudspeaker enclosure or port. Proper tuning enhances bass extension and efficiency. It depends on enclosure volume and port dimensions. Accurate tuning improves system performance.

Turbulent Flow Noise — Noise generated by chaotic, high-velocity airflow. It occurs in ducts, fans, and aerodynamic structures. Turbulent noise has broadband characteristics and can be difficult to control. Smooth flow paths reduce turbulence.

Tweeter — A loudspeaker driver designed to reproduce high frequencies. Tweeters provide clarity, detail, and spatial cues. Their design influences dispersion and tonal character. They are essential in multi-way loudspeaker systems.

UHF — Ultra-High Frequency, the RF band from 300 MHz to 3 GHz. UHF is widely used for wireless microphones, communication, and broadcasting. It offers good range and moderate penetration. Frequency coordination ensures reliable operation.

Ultrasonic — Referring to sound frequencies above the human hearing range, typically above 20 kHz. Ultrasonic waves are used in sensing, cleaning, imaging, and measurement. They exhibit strong directionality and rapid attenuation. Ultrasonics support advanced technologies.

Ultrasonic Frequency — Any frequency above the audible range. Ultrasonic frequencies enable high-resolution imaging and precise sensing. They behave differently from audible sound due to shorter wavelengths. Applications span medicine, industry, and research.

Unbalanced — A signal connection using two conductors, one carrying signal and one ground. Unbalanced lines are more susceptible to noise and interference. They are common in consumer audio and short cable runs. Balanced connections offer better rejection.

Unidirectional Microphone — A microphone designed to capture sound primarily from one direction. Patterns include cardioid, supercardioid, and hypercardioid. Unidirectional mics improve isolation and reduce feedback. Placement is critical for optimal performance.

Unshielded Cable — A cable without conductive shielding around its conductors. Unshielded cables are more vulnerable to noise, interference, and handling artifacts. They are suitable only for low-risk environments. Professional audio relies on shielded designs.

V — A context-dependent symbol that may represent voltage, volume, or velocity. Its meaning depends on the equation or system. Clear notation ensures proper interpretation. V is widely used in acoustics and electronics.

VCA — Voltage-Controlled Amplifier, a device whose gain is controlled by an external voltage. VCAs enable automation, dynamics control, and smooth level changes. They are common in consoles, compressors, and synthesizers. VCAs support precise, repeatable adjustments.

VHF — Very-High Frequency, the RF band from 30 to 300 MHz. VHF offers long-wavelength propagation and good obstacle penetration. It is used in wireless audio, broadcast, and communication. Proper coordination prevents interference.

Vibration — Mechanical oscillation of a structure or object. Vibration can radiate sound, cause noise, or affect equipment performance. Controlling vibration improves comfort and clarity. It is central to structural acoustics.

Vibration Dose Value — A metric that quantifies cumulative exposure to vibration over time. It reflects both magnitude and duration. VDV is used in occupational health and comfort assessment. Higher values indicate greater risk of discomfort or harm.

Vibration Isolation — The reduction of vibration transmission between structures or devices. Isolation uses springs, mounts, pads, or decoupling techniques. It prevents noise, protects equipment, and improves clarity. Effective isolation is essential in studios and buildings.

Vibration Velocity — The speed at which a point on a vibrating surface moves back and forth. It is a key parameter in structural acoustics and sound radiation. Higher vibration velocity generally produces stronger airborne sound. Measuring it helps diagnose noise and vibration problems.

Viscous Losses — Energy losses caused by internal friction within a fluid as sound waves propagate. These losses convert acoustic energy into heat. They are significant at high frequencies and in narrow ducts or pores. Viscous losses influence absorption and attenuation.

Voice Coil — The moving coil in a loudspeaker or headphone driver that interacts with a magnetic field to produce motion. When current flows through the coil, it drives the diaphragm. Voice-coil design affects sensitivity, power handling, and distortion. It is central to transducer performance.

Volt — The unit of electrical potential difference. In audio, voltage determines signal level and gain structure. Proper voltage matching ensures clean, distortion-free operation. The volt is fundamental to all electronic systems.

Volume — The perceived loudness of sound or the physical space of a room, depending on context. In audio systems, volume controls adjust signal amplitude. In acoustics, room volume influences reverberation and modal behavior. Clear context defines its meaning.

Volume of Air — The total amount of air contained within a space. Room volume strongly affects reverberation time and low-frequency behavior. Larger volumes support longer decay and smoother modal distribution. It is a key parameter in room-acoustic design.

VU — Volume Unit, a standardized measure of average audio level. VU readings approximate perceived loudness rather than peak values. They are used in broadcast, recording, and gain staging. VU meters respond slowly, smoothing rapid transients.

VU Meter — A meter that displays average audio level using the VU scale. It helps maintain consistent loudness and prevent overload. VU meters are slower than peak meters, making them ideal for program-level monitoring. They remain common in analog and digital systems.

W — A context-dependent symbol that may represent watt, angular frequency, or wavenumber. Its meaning depends on the equation or system. Clear notation ensures proper interpretation. W appears widely in acoustics and electronics.

Warmth — A tonal quality associated with strong low-mid frequencies and smooth high-frequency response. Warm sound feels rich, full, and pleasant. Excessive warmth can cause muddiness. Engineers shape warmth through EQ, microphone choice, and room acoustics.

Watt — The unit of power, representing one joule per second. In audio, watts describe amplifier output and loudspeaker handling. Higher wattage supports greater loudness but must match speaker ratings. Power specifications guide system design.

Wattage — The power capacity or output of an electrical device. In audio, wattage indicates how much power an amplifier can deliver or a speaker can handle. Proper wattage matching prevents distortion and damage. It influences loudness and headroom.

Wave — A disturbance that transfers energy through a medium without transporting matter. Sound waves are longitudinal involving pressure fluctuations. Wave behavior includes reflection, refraction, and interference. Waves form the basis of acoustics.

Wave Equation — A mathematical equation describing how waves propagate through space and time. It models sound, light, and other wave phenomena. Solutions reveal speed, frequency, and boundary behavior. The wave equation underpins acoustic theory.

Waveform — The shape of a signal when plotted over time. Waveforms reveal amplitude, frequency, and transient characteristics. They help identify distortion, clipping, and noise. Different waveforms produce distinct timbres.

Wavefront — A surface connecting points of equal phase in a propagating wave. Wavefronts may be spherical, planar, or complex depending on the source. Their shape influences propagation and interference. Wavefront analysis supports modeling and design.

Waveguide — A structure that directs and shapes the propagation of sound waves. Waveguides improve dispersion control, efficiency, and directivity. They are used in loudspeakers, horns, and acoustic devices. Proper design enhances clarity and coverage.

Wavelength — The physical distance between repeating points in a wave, such as crest to crest. Wavelength is inversely proportional to frequency. It influences diffraction, interference, and room behavior. Low frequencies have long wavelengths that interact strongly with boundaries.

Wavenumber Spectrum — A representation of wave energy as a function of spatial frequency rather than temporal frequency. It reveals how waves vary across space. Wavenumber analysis supports structural acoustics and modal studies. It complements traditional frequency spectra.

Weighted Decibels — Sound levels adjusted using frequency-weighting filters to reflect human hearing sensitivity. Common weightings include A, C, and Z. Weighted decibels support regulation, measurement, and exposure assessment. They provide more meaningful metrics than unweighted levels.

Weighting — The application of a frequency-dependent filter to sound measurements. Weighting curves mimic human hearing or emphasize specific bands. They are used in SPL meters, noise regulations, and psychoacoustic metrics. Proper weighting ensures relevant results.

Wet Signal — An audio signal that includes effects such as reverb, delay, or processing. Wet signals contrast with dry, unprocessed signals. Blending wet and dry signals shapes depth and texture. Wet/dry balance is central to mixing.

White Noise — A noise signal with equal energy at all frequencies. White noise sounds bright and hiss-like. It is used for testing, masking, and synthesis. Its flat spectrum supports calibration and analysis.

Wiener Filter — An optimal filter that minimizes noise and distortion based on statistical properties of the signal. Wiener filters are used in restoration, noise reduction, and prediction. They require knowledge of signal and noise spectra. Their performance is mathematically optimal under certain conditions.

Window Function — A mathematical function applied to a signal segment to reduce spectral leakage during analysis. Windowing shapes the edges of time-domain data. Different windows balance resolution and leakage. They are essential in FFT-based measurements.

Windscreen — A cover placed over a microphone to reduce wind noise and plosives. Windscreens use foam, fur, or mesh materials. They protect clarity in outdoor and vocal recording. Proper selection depends on wind intensity and microphone type.

Wireless Microphone — A microphone that transmits audio via radio frequency rather than a cable. Wireless systems provide mobility and flexibility. Their performance depends on frequency coordination, antenna design, and RF environment. They are essential in live sound and broadcast.

Woofer — A loudspeaker driver designed to reproduce low-frequency sound. Woofers provide bass foundation and power. Their performance depends on enclosure design, cone material, and motor strength. They are central to full-range loudspeaker systems.

Wow — A slow, periodic variation in pitch caused by mechanical irregularities in analog playback systems. Wow affects turntables, tape machines, and film projectors. It produces audible wavering or instability. Proper maintenance reduces wow.

Wye-Connector — A connector that splits one signal into two paths in a Y-shaped configuration. Wye-connectors are used for simple signal distribution. They may introduce impedance issues if misused. Proper application preserves signal integrity.

XLR — A professional, locking connector used for balanced audio signals. XLR connectors provide noise rejection and secure connection. They are standard for microphones, mixers, and pro audio gear. Their durability supports reliable performance.

X-Y Mixing — A stereo mixing technique that blends two signals using level and panning relationships. X-Y mixing creates stable stereo images with good mono compatibility. It is used in microphone arrays and post-production. Proper balance shapes width and focus.

Z — A context-dependent symbol that may represent impedance, depth, or the third spatial axis. In audio, Z most commonly denotes electrical or acoustic impedance. Clear context ensures correct interpretation. Z appears widely in equations and system models.

Conclusion

Acoustics touches nearly every corner of modern life, from the spaces we inhabit to the technologies we rely on, from the physics of waves to the subtleties of human perception. As the field continues to evolve—driven by advances in digital signal processing, computational modeling, materials science, and auditory research—the vocabulary used to describe its concepts grows just as quickly. Clear, consistent terminology becomes essential not only for specialists, but for anyone working across the boundaries that connect science, engineering, design, and sound.

Acoustics: A Unified Technical Glossary is meant to serve as that foundation. By gathering terms from the full spectrum of acoustic disciplines, it provides a shared language that supports learning, collaboration, and innovation. Whether a reader arrives from physics, audio engineering, architecture, hearing science, or computational acoustics, the glossary offers a way to navigate the field with confidence and precision.

Ultimately, acoustics is a discipline defined by connection—between cause and effect, source and receiver, vibration and perception. A unified glossary strengthens those connections by making the field more accessible and coherent. As new discoveries emerge and technologies advance, the language of acoustics will continue to grow, and this glossary stands as a resource to support that ongoing exploration.

www.ingramcontent.com/pod-product-compliance
Lightning Source LLC
LaVergne TN
LVHW011521200926
848484LV00001B/94
* 9 7 9 8 2 9 5 5 4 8 8 8 8 *

